UNIVERSITY CHEMISTRY SERIES

Editors:

(*Organic*) Professor M. F. Grundon, M.A., D.Phil. (Oxon) New University of Ulster.

(*Inorganic*) B. C. Smith, M.A. (Cantab.), Ph.D. (Nott.). D.Sc. (Lond.) Birkbeck College, University of London

(*Physical*) Professor M. W. Roberts, B.Sc., Ph.D. (Wales) Bradford University

CHEMICAL THERMODYNAMICS
WITH SPECIAL REFERENCE TO INORGANIC CHEMISTRY

CHEMICAL THERMODYNAMICS

WITH SPECIAL REFERENCE TO INORGANIC CHEMISTRY

D. J. G. IVES

Professor of Electrochemistry
Birkbeck College
University of London

MACDONALD : LONDON

© *D. J. G. Ives*, 1971

First published in 1971 by

Macdonald & Co. (Publishers) Ltd.

49–50 Poland Street

London W.1

ISBN 0356 03736 3   hardback
ISBN 0356 03871 8   paperback

PRINTED BY W. & G. BAIRD, BELFAST

Contents

Preface

An addition to the many existing books on thermodynamics can be justified only by special objectives. The writer has three in mind.

(i) He associates himself with unfairly treated students who, under the load of an ever-expanding curriculum, are expected to digest in a few months what took the Masters many years of thought to develop. It is an ill-kept secret that they do not all succeed. The first aim is, therefore, to come to the assistance of students in meeting the not very reasonable demands that face them. Accordingly, the writing is from ground level, with illustrations and applications suited to tie the subject firmly to the realities of chemistry.

The strategy is firstly to describe the situation as the writer sees it, and to identify the attitudes and disciplines of mind specially needed for the study of thermodynamics. This is the concern of chapter 1, presented without apologies. Secondly, because a student should know something of what a subject is about and what are its powers before he can be expected to devote mental exertion to it, chapter 2 gives a broad picture of the role of thermodynamics in chemistry. These two chapters have no sub-headings because they are meant to be read as essays. In the following chapters, the writer gets down to brass tacks, without losing sight of the first objective.

(ii) The second end in view is to provide a text on the applications of thermodynamics in inorganic chemistry acceptable to advanced students. In a short book, the writing is necessarily selective, and there can be no pretence to cover the field. If examples other than inorganic best suit the development of a thermodynamic theme, they are used. The main purpose throughout is the clarification of principles. Chapters 3 to 6 are sub-divided into appropriate sections, so that the course of development can be seen in advance and individual sections can be consulted.

(iii) The third objective was to assist in demolishing the artificial barrier that has grown up between 'pure' and 'applied' branches of science, with an absurd and damaging inferiority of prestige accorded to the latter. Unfortunately, this objective has not been attained. Time and other adverse circumstances have overtaken the writer, and this aspect, with other matters, must be left to a future effort.

Throughout, pains have been taken to make the book readable. No background of thermodynamics is assumed, little of mathematical language, but quite a lot of chemistry. References of two kinds are given: to original papers, and to texts for general reading. These can be distinguished by the context in which they are given. In quoting numerical data, the writer has not used the so-called S.I. units, nor has he adopted recent recommendations on symbols—already sufficiently publicised.

This is because he is writing for a generation of students that will have to read the scientific literature and be familiar with long established usages, which the new proposals disregard. Whether the proposals will be accepted or rejected is not yet clear; it must ultimately depend on their merits. As it stands, the writer, with many others, does not like them, disapproves of regimentation, and cannot agree that a uniform system of units and symbols throughout science is practicable or desirable. He must not, however, press his own prejudices and therefore appends to this Preface the only conversion factors between new and old units that readers will need.

If it turns out that the author has written a worthwhile text, it may be, in part, thanks to the self-confidence engendered by the compulsion to think thermodynamically about his own work. Perhaps latent in this remark is the primary message to students. Turning the corner between learning about a subject and making it one's own is largely a matter of self-confidence to be gained not only by private study but by argument and discussion with one's peers. Courage is close to self-confidence, and calls for a degree of aggressiveness. The student who sits down saying to himself 'Now what the hell is all this nonsense about?' is on the right lines. Thanks will also be due to the author's students, who have posed questions, to his colleagues and his teachers.

Gratitude is especially due to Dr. B. C. Smith for his help and encouragement.

January 1970   David J. G. Ives

CONVERSION FACTORS

1 thermochemical calorie $= 4 \cdot 184$ J

1 electron volt $= 1 \cdot 6021 \times 10^{-19}$ J

1 atmosphere $= 101 \cdot 325$ kN m^{-2}

1

Introduction

Having agreed to prepare this book, the writer faced problems peculiar to the subject. The nature of these problems, and the ways chosen to resolve them, determined the nature of the writer's effort, and must be declared to potential readers.

The commission was, to address the work primarily to final-year under-graduate students. Accordingly, the first step was to size up these students and assess their requirements. Ideally, the assumptions would be that they have mastered their earlier studies, that they have already got (or will very soon get) a grasp of basic thermodynamic principles, and that they will securely retain in after-years what they have learned about the subject. Realistically, these assumptions are questionable. As an undergraduate, the writer's thermodynamic know-how (despite excellent instruction) was formalised and superficial. He believes that he was not peculiar in this respect, nor would be so in an average company of present-day students. If this is a true reading of the situation, the aim can be widened to take in first- and second-year students, who may welcome a short text somewhat unconventional in approach, emphasis and illustration. If the book is a help to them, future final-year students will also be served. Another corollary to the argument is that there will be many postgraduates, various in seniority and occupation, who may be glad to refurbish and extend their knowledge—they are, after all, sometime final-year students.

It may seem odd for an author to begin by offending his student readers by casting aspersions on their grasp of the basic principles of the subject about to be discussed and developed to a reasonably advanced level, but to be candid is the first necessity. If the aspersions are not unjustified, the question arises, are there any excuses, built-in to the history and nature of the subject, that the students could plead? There are indeed, and they should be exposed, as follows.

A greenhorn freshman, finding 'thermodynamics' listed in his chemistry curriculum, might be excused for going to his dictionary. Finding 'the science of the relation between heat and mechanical energy', he could suppose that there had been a mistake. His ancillary physics lectures would soon dispel this idea. Later, he might well find himself being led round a 'Carnot cycle' in his own department, but (notwithstanding its fundamental implications) this classical theorem now induces an anaphylactic reaction; paralytic boredom sets in, and the damage is done. But tradition dies hard; this is how thermodynamics began—let us not play fast and loose with the pure doctrine. The student, no longer so green, sees a way out. After all, thermodynamics is but a sub-section of physical chemistry, which is in turn a sub-section of chemistry, so thermodynamics is only a sub-sub-section of chemistry, limited in scope and restricted in interest. Why bother with it? There is a superfluity of more interesting and recent material to study, and putting his back into this will get him a good degree. So it does—nor undeservedly. But he has been served ill if he carries away the impression that there is anything 'sub-sub' about thermodynamics, or, indeed, that science is as compartmentalised as the 'lectures-cum-examinations' system tends to suggest. Perhaps, because of pressure of time, there is a general failure to show students how all the fields of science interpenetrate and adapt to their own purposes the same fundamental angles of study—a failure even to the extent that a chemistry student may be left unaware that thermodynamics is indispensable in engineering, metallurgy and biology, let alone in his own subject.

The student can certainly blame 'the squeeze of learning'. Most degree courses take no longer than they did thirty years ago, but now contain a great number of things not then heard of. The total of published science now doubles itself in about seven years; the load is great and increasing, and we must keep up-to-date. The student (no longer monkish) has interests outside his studies. Time is at a premium, with little to spare for an intellectually demanding subject, not to be mastered sheerly by learning—an old subject, hag-ridden by tradition, at that.

Is thermodynamics (personified) blameworthy in other respects for its somewhat repellent image in students' eyes? Perhaps it is, tending to be inward-looking, Narcissus-like in contemplation of its own intellectual beauty. Texts of unassailable authority develop the subject sequentially with pitiless rigour—of inestimable value, these, but too strong meat for the tyro. Respect bordering on awe is demanded for the immutable laws of thermodynamics. Yet there might be some difficulty in rebutting an unimpressed student's assertion that the first, second and third laws are, respectively, but a definition, a stumbling-block and a pious hope.

To adopt an even more aggressive posture of attack (protest is in fashion), it could be said that there is still often too much mystique in presentation of the subject; too much formalism; too much going round in cycles to prove the immediately acceptable, and too much reluctance to discard outdated devices (e.g. the equilibrium box) in favour of the much shorter and clearer lines of thought for which Willard Gibbs is mainly to be thanked. Too little is done to dispel a particular aspect of the 'closed shop' image of thermodynamics. This is, that it is never able to give any information of interest outside its own restricted field—at best, it throws a spanner in the works of somebody's bright idea by showing it to be thermodynamically inadmissible. The trouble is, this allegation is a half-truth—but the other half is violently false. Thermodynamics has something fundamental to say in so many fields. Apart from the service of laying down the conditions under which a bright idea may be expected to work, thermodynamic data frequently carry a message—'here is something of exceptional interest to be investigated by all appropriate means: spectroscopy, ultrasonics, neutron scattering, all the relevant complementary angles of study'. A scientist unable to appreciate or use the thermodynamic angle of approach to natural problems has a disability; illuminating avenues of thought are closed to him.

After the preamble on the nature of the subject, it is logical to take yet another look at the student. What qualities are required of him for success in his studies of *this* subject in particular?

The general requirement is reasonable facility and habitual care in the use of two languages. The first of these is the student's native tongue (or other language in which he must pursue his studies). It is here that the first difficulties arise. No, the writer is not about to affront his readers again (unless the cap fits), but merely wishes to remark that English, especially, is a very flexible language. Words vary in meaning with the context in which they are used, or even by whom. Familiar words in daily use, such as 'system', 'equilibrium', 'reversible' and 'probability', are also used in science, but they must now each serve the purpose of conveying an invariable and unambiguous meaning—one that requires careful and particular definition. Once the definition has been made, the word must be the master.

The second essential language is mathematics. By the authority of Willard Gibbs, 'Mathematics is a language'.*

A verbal statement of a relationship between physical quantities clearly has its analogue in the form of an equation, using symbols instead of words. Provided the symbols are defined in meaning, the equation is more concise

---

* To quote from G. N. Lewis and M. Randall, '*Thermodynamics*', McGraw-Hill, New York, 1923 (a famous book revised by K. S. Pitzer and L. Brewer in 1961): 'It is said that, during his long membership in the Yale faculty, Willard Gibbs made but one speech, and that of the shortest. After a prolonged discussion of the relative merits of language and mathematics as elementary disciplines, he rose to remark "Mathematics is a language." '

and lends itself better to reasoning processes. Sight must not be lost of the two-way nature of the relationship between the two languages. Every equation is saying something, and its translation into words with an intelligible physical meaning is an essential exercise—within reasonable limits. The mathematical language is nimble; the verbal equivalent of even a short series of mathematical operations would be cumbersome, tedious and ultimately pointless, given grasp of its mathematical counterpart.

It is here that the second crop of difficulties arises—from students' lack of the right kind of mathematical background. Even those not so hampered may have initial difficulties. For instance, a theorem may start from a simple equation with a clearly identifiable physical meaning. It is subjected to a sequence of mathematical operations in which the obvious link with reality is lost, and eventually gives rise to an equation of unanticipated form— to be regarded with healthy distrust. Does it mean anything real? It must be translated to see what it does mean, and it must then be tested against experimental facts. When it is found that what the final equation is saying is true, and represents an illuminating and useful result of valid reasoning, confidence is engendered. Examples follow in due course.

Appraisal of the writer is a logical sequel to the assessments of subject and students. This is up to the readers in due course, but some pre-emptive admissions can be made. He has an innate stupidity adequate to appreciate students' difficulties. He has been in circulation long enough to see through most of them, and too long to be deceived into believing that they are not widespread. At the same time, he is happy to admit that his possibly overly subjective and gloomy comments will be found outdated and inappropriate in some quarters. It is, however, the students who *are* in difficulties that are his main concern, and his aim is to make as accessible to them as to the natural-born eggheads the matters of interest offered by chemical thermodynamics.

The writer denies being stupid in thinking that students have a right to be shown something of what a subject is about, and why it is important, before being called upon to exert considerable efforts in studying it. This is the idea behind Chapter 2. For the rest, the emphasis is mainly, but not over-overwhelmingly, on the thermodynamics rather than the inorganic chemistry. The right has been exercised to use any illustrations, inorganic or not, as best serve the main purpose. Since the book is short, it must be limited in scope and ancillary in nature. Nevertheless, the writer has tried to make it self-sufficient, not recoiling from ultimately elementary considerations. 'Basic' is a better word than 'elementary', and it is often in such matters that students, unaware, come unstuck. Most of us have a few little things to be shame-faced about.

2

Chemical reactivity

A concept so important as chemical reactivity must be quantified. That thermodynamics is essential but insufficient for this purpose can be illustrated by an example.

Consider two gas mixtures. The first is of nitrogen with half its volume of oxygen. The second is similar, except that carbon monoxide is substituted for nitrogen. Both can be kept indefinitely, without change; both conform to the gas laws. They are, indeed, very nearly physically indistinguishable, and it would be reasonable to describe them both as 'systems in equilibrium', with properties uniquely determined when the temperature and pressure have been fixed. The effects of momentary exposure to a hot wire, or of contact with a catalyst, are, however, different. The first mixture ($N_2 + \frac{1}{2}O_2$) suffers no permanent change. The second mixture ($CO + \frac{1}{2}O_2$) is triggered into exothermic chemical reaction, and is never the same again.

The first implication is that the expression 'in equilibrium' needs qualifying—at least by 'physical' or 'chemical', but the immediate problem is as follows. The second mixture had a natural tendency to undergo a chemical reaction, releasing energy and reaching a more stable final state. Why did this reaction not take place as soon as the gases were mixed? The 'motive power' was there, so there must have been some sort of barrier to reaction, overcome only by provision of a local hot-spot or a catalyst. Before the catalyst was added to the mixture, the rate of reaction was zero; afterwards, it was great.

This is but a demonstration (at 'basic' level) that two questions must be asked about any chemical reaction—or, indeed, about any process. These questions can be framed, using everyday language, in alternative but related ways, as follows:

1. Has the reaction its own motive power or driving force?
   Is the reaction capable of occurring spontaneously?
   Is the reaction accompanied by an increase in the stability of the system in which it takes place?

Does the reaction represent an approach to an equilibrium state?
Can the reaction conceivably be harnessed to provide some useful work?
2. How fast does the reaction go?
Is the reaction opposed by a barrier to reaction?

These questions, in all their forms, clearly have their more exact and quantitative counterparts. If they can be fully answered for numbers of chemical reactions, progress is being made towards quantifying chemical reactivity. Since it would be absurd to ask how fast a reaction will take place before it is known whether it can go at all, the questions must be asked in the correct order. For instance, the answer to question 1 for the first gas mixture would (under normal conditions) be *no*, so question 2 would not arise.

Question 1 is the subject matter of thermodynamics; question 2 is the subject matter of reaction kinetics.

A start must therefore be made from the thermodynamic angle, perhaps by extending question 1 into 'How do we measure the driving force of a reaction?'

This question must be examined. Its meaning is relatively clear, but phrases like 'motive power' and 'driving force', although expressive, will not do because they are inexact. Force and power are not the same thing, and neither is really intelligible in relation to a chemical reaction. A word not misleading because of its other connotations, is needed. The agreed word is *affinity*. Then, reframed, the question is 'How do we measure the affinity of a reaction?' and it means 'How do we measure the *thermodynamic tendency to occur* of a reaction?', on the clear understanding that questions about reaction rates are separate, and come later.

A next step might be to ask 'What factors contribute to the affinity of a reaction?' But does it not seem likely that each of the many thousands of known reactions would require its own particular answer? This is indeed the case, but it is sensible to look first for general principles—unifying generalisations which override all the complications. It is the general principles that are the first concern of thermodynamics, which should therefore be seen as the co-ordinating and simplifying science. Mechanistic details are studied by other methods, reaction kinetics being one of the most fruitful.

It is becoming clear that a basic necessity is to be able to ask unambiguous questions. Another is to know what sort of answers to expect. In other words, there is a basic language to be learned, and this is where the writer dare not skip anything.

Thermodynamics is largely, but not exclusively, concerned with energy. The various kinds of energy—mechanical, electrical, radiant, chemical, thermal—are interconvertible without loss. This is implicit in the *first law of thermodynamics*—the law of conservation of energy, which states that energy can neither be created nor destroyed. What is the positive proof of this? There is none. It is a scientific belief which no experience has contradicted, provided that mass is included as a form of energy in terms of the Einstein relation, $E = mc^2$, which is now common knowledge. Nuclear chemistry is, but chemical thermodynamics is not, concerned with mass/energy interconversion. The classical first law remains the basis of thermodynamic reasoning.

All forms of energy have the same *dimensions*. The general meaning of this statement is to be explained as follows.

Many of the measurable physical quantities coming into scientific consideration can be seen to be derived from but a few more basic quantities—usually, but not uniquely, mass, m; length, l; time, t; temperature, θ. Thus:

Velocity = length/time lt^{-1}

Momentum = mass × velocity mlt^{-1}

Acceleration = rate of change of velocity; velocity/time lt^{-2}

Force; measured in terms of the acceleration imparted to a mass
    = mass × acceleration mlt^{-2}

Energy; measured in terms of the work it will perform; work
    = force × distance ml^2t^{-2}

Power = rate of performance of work ml^2t^{-3}

Heat capacity = rate of increase of heat content with temperature
    $ml^2t^{-2}\theta^{-1}$

These considerations are included because attention to the dimensions of quantities is as essential in thermodynamics as in any other field of science. Only quantities of the same dimensions can be equated to each other, or added to, or subtracted from, each other. Any dimensionally inconsistent equation is a wrong equation.

Energy, then, and the work it may perform, has the dimensions ml^2t^{-2}. <u>Units</u> are a separate matter, and must be appropriately chosen. Clearly, within a given calculation, there must be consistency of choice. There is no fundamental difficulty because the conversion factors relating amounts of different kinds of energy are invariant natural constants, known with progressively increasing accuracy. All that is normally required in thermodynamics is

10^7 ergs = 1 joule, watt-second, volt-coulomb
4·1840 joules = 1 calorie

Thermodynamics is also concerned with matter *in bulk*. The *mole* is the appropriate unit of quantity of chemical substances, not the molecule.

Since there are $6 \cdot 0225 \times 10^{23}$ molecules in a mole, thermodynamics deals with vast assemblies. This is more than just important—it is essential to the nature of the subject. All thermodynamic systems must be *macroscopic*—as opposed to microscopic.

Not skipping anything connected with language, the word *system* requires attention. It means the quantity of matter, collection of materials—anything from a fuel cell to a tin of soup that comes into thermodynamic discussion. The first necessity is to define it and distinguish it from its *surroundings*. Failure to make this distinction can, and does, lead to difficulties. Most systems of interest are in thermal and mechanical contact with their surroundings. If, however, a system is thermally insulated from its surroundings, it is *adiabatic*; if it is insulated against transfer of any kind of energy, it is said to be *isolated*. Systems constant in material content are called *closed systems*—in contrast to *open systems*, for which material content is treated as a controllable, independent variable. Pure thermodynamics deals with systems not undergoing net change; systems that are at least in relative equilibrium states, stable with respect to all slightly different states. It is clear that some aspects of this concise statement need (and will later get) further discussion—the language is not as simple as might be supposed, and calls for all the closer attention.

Considering only closed systems, the questions arise 'How is a system to be described? How many statements must be made to provide an unambiguous definition of the system and of its *state*?'

First, the material content must be specified; what substances, how much? Then temperature and pressure must be stated, and, normally, that is all. What about the volume? This need not be stated, because the system, once adequately defined, could have but one, unique volume at the temperature and pressure already quoted. Only two *independent variables of state* are normally required to fix the state of a given, defined closed system. These are usually, but not necessarily, temperature and pressure; when these are fixed, all the other properties of the given system are also fixed by the laws of nature, as they apply to that system in its equilibrium state.

Two kinds of property can be distinguished in thermodynamic discussion of systems. Consider a simple system consisting of pure water. One set of properties to be listed would depend on how much water the system contained. These are called the *extensive properties*, because their values depend on extent, magnitude or quantity. The value of an extensive property of a whole system is the sum of the like values for the parts of the system. Such properties are mass, volume, heat capacity and total energy content. Another set of properties would be independent of (in the example) how much water was present; quantity would be irrelevant—as long as the system remained macroscopic. These might be temperature, pressure, specific heat, refractive index—all *intensive properties*, characteristic of 'quality' and defining state. It is easy to understand that the intensive properties of half a mole of water may be the same as the like properties of a mole of water, but this is not the case for any extensive property.

At this point, the reader may well be getting restive at what he regards as ineffably tedious labouring of the obvious. This is one of the pitfalls. The concept of extensive property seems so simple that there is an inclination to dismiss it as self-evident, and not worthy of close attention. Yet it lies at the root of thermodynamic thought. The statement that energy content is an extensive property of a material system tacitly incorporates the first law.

The energy content of a system, however, is not unambiguously determinable on any absolute scale common to all systems, for want of an agreed zero. Thermodynamics, unconcerned, recognises that each defined system, in a state fixed by specifying a sufficient number of independent variables of state, has an extensive property of the nature of total energy. It is called *enthalpy*, is symbolised H, and has dimensions ml^2t^{-2}. If a system in state 1, with enthalpy H_1, is transformed, perhaps by chemical reaction, into state 2, with enthalpy H_2, then the change in enthalpy, ΔH is

$$\Delta H = H_2 - H_1 \qquad\qquad (2.1)$$

If $H_1 > H_2$, ΔH is negative, and the transformation is accompanied by loss of enthalpy.

If $H_2 > H_1$, ΔH is positive, and the transformation is accompanied by gain of enthalpy.

More blinding glimpses of the obvious. Obvious or not, equation (2.1) is not only basic (incorporating, or derived from, the first law), but misunderstanding of it is a not uncommon source of error.

It is, in principle, easy to carry out reactions in a calorimeter at constant pressure, ensuring that the lost enthalpy appears only as heat, or the gained enthalpy is absorbed only as heat. In other words, although H_1 and H_2 cannot individually be measured, the difference between them, ΔH, *can* be— not only by calorimetry, but in other ways as well.

Consider an example.

Let state 1 of a system be

$$\begin{cases} 1 \text{ mole of } H_2 \text{ gas at 25 °C and 1 atm pressure} \\ \tfrac{1}{2} \text{ mole of } O_2 \text{ gas at 25 °C and 1 atm pressure} \end{cases}$$

and state 2,

1 mole of liquid H_2O at 25 °C and 1 atm pressure.

By calorimetry and other methods, it is found that

$$\Delta H = -68 \cdot 32 \text{ kcal}$$

Since when enthalpy is lost (ΔH negative) heat is liberated, and because the states of reactants and product have been adjusted to the arbitrary but agreed standard values of 25 °C and 1 atm, the result may be written

$$\Delta H^\circ = -68 \cdot 32 \text{ kcal}$$

and is called the standard enthalpy of formation of water.* The superscript symbol conventionally denotes that all the participants in the chemical transformation concerned are in *standard states*, open to arbitrary choice and requiring definition. The agreed choice for the extensive tabulations of standard enthalpies of formation of compounds from their constituent elements are the normal states at 25 °C and 1 atm pressure.* Discussion of the use of such standard data is deferred.

At this stage, chemists will feel a compulsion to look at ΔH for chemical reaction from a characteristically *chemical* point of view, as follows.

Chemists find it intellectually acceptable (they were brought up on it) that atoms and molecules seek to attain the most stable electronic configurations available to them. In the processes by which this tendency is satisfied, electron transfer or sharing may occur, but however complex the rearrangements concerned, they are always such as to lead to a fall in potential energy. In the most general terms, when forces of attraction of any kind are satisfied, there is a fall of potential energy, and, since energy is conserved, a liberation of heat. If chemical bonds which did not exist before can be made, or if stronger bonds can be formed at the expense of breaking weaker ones, or if any forces of attraction—as between ions of dissimilar sign of charge, or van der Waals forces between molecules—are spontaneously satisfied, ΔH for the process concerned would be expected to be negative.

It might be thought, then, that we have in $-\Delta H$ the measure of affinity, or thermodynamic tendency to occur, that is needed.

There is, however, something wrong with this argument, however reasonable—indeed compelling—it seems. The existence of spontaneous *endothermic* reactions leaves no doubt about this. These reactions proceed of their own accord, absorbing heat, ΔH being positive. The products lie at a higher energy level than the reactants. What factor is it that can push a system up an energy hill, and what is the fallacy in the above 'chemists' viewpoint'?

These questions (following so quickly on 'glimpses of the obvious') mark

* There is a difficulty of language and convention associated with this statement; it is too important for a footnote and is discussed in the opening of the next chapter.

the first arrival at the nub of thermodynamics. Full answers require feedback from further studies, but a start can be made along the following lines.

Chemical bond formation is considered on the atomic-molecular scale, in terms of individual events in microscopic systems. For instance, an objective of theoretical-spectroscopic studies of the hydrogen molecule is to find how much lower the energy of *one* molecule lies than that of *two* separated hydrogen atoms. The energy difference is a valid measure of the affinity of the reaction $2H \rightarrow H_2$ as it might be imagined to occur between two hydrogen atoms in the isolation of free space at $0\ °K$. The result is important, but divorced from practical affairs. It is a fact that two hydrogen atoms will *not* combine under such conditions, because the energy released would be more than the molecule could accommodate—it would be raised to a temperature high enough to dissociate it into atoms. If combination is to be effected, a 'third body' is necessary to carry away the excess of energy.

What would be the thermodynamic view of this reaction? What additional factors would, in general, have to be taken into account to establish the link with the realities of natural processes occurring in macroscopic systems?

It is essential to recall that thermodynamics deals *only* with macroscopic systems. There is no shortage of 'third bodies'. There is a means for the distribution of thermal energy and for the liberation of heat to the surroundings. These processes are part and parcel of the 'thermodynamic reaction'. Thus, for the example in question, the normal thermodynamic treatment would be in terms of the transformation

2H

(2 moles of atomic hydrogen gas
at 25 °C and 1 atm pressure)

↓

H_2

(1 mole of molecular hydrogen gas
at 25 °C and 1 atm pressure)

Quotation of initial and final standard states* invariably presupposes the existence of physical equilibrium. In this case, the thermal energy of the atomic hydrogen (translational) must be appropriate to the stated temperature, and similarly for the thermal energy of the molecular hydrogen (translation, rotation, vibration). Except for some give and take consistent with these requirements, all the energy liberated in bond-forming is got rid of to the surroundings.

---

* Standard states, open to choice, do not have to be physically realisable, and often are not. It might be more appropriate to choose 1500 °C and 10^{-5} atm in this case, but the argument is not affected.

This leads directly to a general observation of importance. The heat evolved when a chemical reaction occurs in a macroscopic system at constant temperature and pressure has its origin in the large energy release of the quantum-mechanical processes of bond-making, but these two energy terms cannot be equated for two reasons. The first is that there is a contribution to the enthalpy of a system of finite volume V under pressure P equal to the product, PV. Change in volume at constant temperature and pressure therefore makes a contribution to the ΔH of reaction; if there is a contraction, there is an increment to heat evolved. The effect is very small for reactions not involving gases; for those that do, it is easily calculable by an obvious application of the general gas law, $PV = nRT$.

The second reason is less obvious. It is because there is generally a difference between the total *heat capacities* of the products and the reactants—that is to say, their requirements of thermal energy to 'fill them up' to the same temperature are not the same. If, then, the reaction is accompanied by a net increase in the heat capacity of the system in which it occurs, the heat evolved will be less than the total energy available from the bond-making. If it is the other way round, some of the thermal energy originally contained in the reactants will 'spill over' and augment the heat evolution. It is an obvious corollary that heat of reaction will vary with the temperature at which the reaction takes place.

All this, however, is but tangential to the main question of why enthalpy loss $(-\Delta H)$ is not a valid measure of affinity. What other factor is significant?

A 'first answer' (pending feedback) can be approached by considering a simple, non-reacting system in thermal and mechanical equilibrium, such as a mole of a gas. If the temperature and the pressure are fixed, the volume and total energy are automatically fixed also, and the *macrostate*, or thermodynamic state, of the system is defined. But the total energy is made up of the kinetic energies of *random* thermal motion of the immense number of molecules that the system contains. Viewed on the molecular scale, it can be seen that there must be a variety of ways in which the energy can be shared out. For example:

Each molecule might have the same energy, precisely equal to the total energy divided by the number of molecules present ($\approx 6 \times 10^{23}$).

Half the molecules might have twice this energy and the other half none.

Very many such possibilities suggest themselves. Each possibility will correspond to an instantaneous state of the system, or *microstate*. It might indeed be thought that there is an *infinity* of microstates, all consistent with the same total energy. This is not so because of what Hinshelwood* has

---

* In his book, *The Structure of Physical Chemistry*, to which the writer owes much. It remains a great gift to the mature student wishing to consolidate his philosophical appreciation of the subject.

called 'the control of the chaos by the quantum laws'. The energies of atomic and molecular units are not continuously variable, but exist in discrete quantum levels. This means that there is a finite, if very large, number of microstates possible to the molecular assembly of defined total energy. It is of fundamental significance that all these microstates have an equal chance of occurrence; all may be regarded as 'impartially explored' by the system in course of time, provided that the system is in equilibrium. This is true even of the two 'unlikely' microstates mentioned above—their chance of occurrence, with all the rest, is of the order of one in something like 10^{25}. It is in this 'impartiality' that the real significance of *randomness* in relation to molecular motion and energy distribution is to be found.

Once the parts played in this problem by the quantum laws and by statistics have been seen, it is best to invert the original point of view and think of how the molecules of the assembly are to be distributed over the available energy states (rather than the other way round). It then becomes clearer that a macrostate offering a larger number of microstates (all to be explored) than another is the more probable. This at least gives a clue to the basis of reasoning behind the following bald statements—all that is permitted by the limited scope of the present discussion.

The *thermodynamic probability*, W, of a macrostate of a system is equal to the number of microstates that contribute to it, or, what is the same thing, the number of ways in which the macrostate can be realised. An equilibrium state can be realised in a greater number of ways than any other. If the state of a system is the equilibrium state, W is at a maximum; if it is not, W is less than the maximum. Clearly W (an absolute probability, not a relative one) is a very large number. The function of W which is of the nature of an extensive property of a system is the *entropy*, S.

$$S = k \ln W \tag{2.2}$$

where k is the Boltzmann constant, $1 \cdot 3805 \times 10^{-16}$ erg $°K^{-1}$ molecule^{-1} ($= R/N$, where R is the gas constant and N is the Avogadro number).

Entropy, then, is an extensive property of a system associated with probability and randomness. It has the dimensions $ml^2t^{-2}\theta^{-1}$ and is measured in cal $°K^{-1}$. Unlike enthalpy, entropy can be measured on an absolute scale; there are extensive tabulations, constantly enlarged, of the standard entropies, $S°$, of pure substances (cal $°K^{-1}$ mole^{-1}) at 25 °C and 1 atm pressure. Two such values are:

diamond, $0 \cdot 58$ cal $°K^{-1}$(g atom)$^{-1}$
neon, $34 \cdot 95$ cal $°K^{-1}$ mole^{-1}

The first of these substances is a very hard and strong, perfectly ordered, covalent crystal—a giant molecule. An assembly of Avogadro's number of carbon atoms in this form has an extremely low entropy. The second is a completely disordered, structureless gas; at the same temperature, it has a high entropy. In general, entropy increases with increasing mass, molecular complexity, temperature and with increasing disorder and mixed-upness of all kinds. An entropy jump accompanies melting, and a much larger one accompanies evaporation. Entropy is a function which is not at first easily appreciated. There is, however, a sound basis for thinking of it in terms of randomness and probability, and understanding comes with use. It is the other factor which must be set beside energy in assessing affinity.

A summarising statement can now be made.

A system in *dis*equilibrium has a natural tendency to undergo a spontaneous process of change until equilibrium is reached. In principle, any such spontaneous process can be made to supply useful energy or work. The amount of work obtainable measures the extent of the initial disequilibrium, and the tendency to occur, or the affinity, of the process that relieves it.

A system may be in disequilibrium for either or both of two reasons. It may be in a state of more than minimum energy, or it may be in a state of less than maximum entropy or probability.

This has to be put into mathematical language.

The enthalpy of a system, thought of as its total energy, can be divided into two parts, both, of course, with the dimensions of energy:

$$H = G + TS \tag{2.3}$$

where T is the absolute temperature. When such equations are written, they are always, if tacitly, making a statement about a system in equilibrium. It is therefore understood that the entropy, S, of the system has the maximum value appropriate to the equilibrium state at the temperature T. All the molecules within the system must have the maximum randomness of spatial distribution and thermal motion proper to this temperature, within the restrictions imposed by the forces that operate, and the structures that exist, within the system. TS thus represents the energy which the system must have, by its very definition, to be in its equilibrium state at the temperature T. TS can therefore be regarded as 'unavailable energy', so that G, by difference, can be called 'free energy'.

It is recalled that thermodynamics is mainly concerned with processes of change from one equilibrium state to another. Since H, G and S are all extensive properties, then for a change, from an initial state 1 to a final state 2, $\Delta H = H_2 - H_1$; $\Delta G = G_2 - G_1$; $\Delta S = S_2 - S_1$, so that for any change at the constant temperature T

$$\Delta G = \Delta H - T\Delta S \tag{2.4}$$

A negative value for ΔG corresponds to a loss in the free energy of the system as a result of the change. It is the free energy *loss*, $-\Delta G$, which is equal to the useful energy or work that can be gained from the change, and is therefore the measure of the affinity of the change, i.e. of its natural, thermodynamic tendency to occur.

Keeping equation (2.4) in view, it is clear that ΔG, the full title of which is the *Gibbs free energy change* for a process, can be negative for either or both of two reasons. If, during the change, process or reaction concerned, the energy content of the system can fall, this will make ΔH negative. In turn, this will make ΔG more negative (or less positive) by an equal amount. If the entropy content of the system can rise, ΔS will be positive, and this will make ΔG more negative to an extent which increases with rising temperature.

Whether the two terms on the right-hand side of equation (2.4) support or oppose each other, it is ΔG that gives the significant result. It may be noted now, although the proof must be deferred, that the temperature coefficient of the affinity $(-\Delta G)$ is equal to ΔS. This means that if a process or reaction is attended by an increase in entropy, it is progressively favoured by rise of temperature.

At this stage, it is desirable to pause, sit back and take a wider view. A point has been reached where it is possible to discern the two main influences of nature which shape things as they are. First, there are the forces of attraction, chemical or physical, tending to produce order and structure by bringing together the elementary building units of matter. As these forces are satisfied, potential energy falls. Secondly, there are the randomising effects of the maximisation of probability or entropy, promoted by thermal motions. The lower the temperature, the greater the play the constructive forces have; the higher the temperature, the greater the destructive effects of what may be called 'blind statistics'. The interplay of these trends will be a running theme throughout this book.

It is instructive to examine the balance of energy and entropy effects in chemical reactions, for there is wide variation from case to case. Two examples must suffice for the present.

$$H_2(g) + \tfrac{1}{2}O_2(g) = H_2O(l) \text{ at } 25\,^\circ C \text{ and } 1 \text{ atm}$$

$$\Delta H^\circ = -68 \cdot 3 \text{ kcal}; \ \Delta S^\circ = -39 \text{ cal }^\circ K^{-1}; \ T\Delta S^\circ = -11 \cdot 7 \text{ kcal}$$

therefore

$$\Delta G^\circ = \Delta H^\circ - T\Delta S^\circ = -56 \cdot 6 \text{ kcal*}$$

* When ΔX (where X is any extensive property) is quoted for a *reaction* duly specified by means of an equation, it should not be quoted 'per mole'; it relates to the whole system undergoing the reaction.

ΔG° is negative because ΔH° is large and negative. This is due to the substantial energy release in net bond-making (number and strength). The entropy change, ΔS°, is negative, as to be expected for the replacement of $1\frac{1}{2}$ moles of high-entropy gases by one mole of much lower entropy liquid. This is unfavourable, and opposes the 'energy drive'. This is why the affinity of the reaction, $-\Delta G^\circ$, is *less* than would be expected from the energy loss, $-\Delta H^\circ$. It can be seen at a glance from the negative ΔS° that the affinity of this reaction will decrease with rising temperature.

$$C(\text{graphite}) + \tfrac{1}{2}O_2(g) = CO(g) \text{ at } 25\,^\circ C \text{ and } 1 \text{ atm}$$
$$\Delta H^\circ = -26\cdot5 \text{ kcal}; \ \Delta S^\circ = +21\cdot5 \text{ cal }^\circ K^{-1}; \ T\Delta S^\circ = 6\cdot4 \text{ kcal}$$

therefore

$$\Delta G^\circ = \Delta H^\circ - T\Delta S^\circ = -32\cdot9 \text{ kcal}$$

Here, ΔS° is positive. A mole of low-entropy solid and *half* a mole of high-entropy gas is replaced by one mole of high-entropy gas. The entropy effect supports the energy effect, and the affinity is greater than would be expected from the energy loss alone. At a glance, it is seen that the affinity of this reaction increases with rising temperature—a fortunate circumstance on which the whole of modern civilisation depends.

It should now be obvious that all spontaneous endothermic reactions are 'entropy driven'.

If it can now be tentatively agreed (pending feedback) that the affinity, or 'thermodynamic tendency to occur', of a reaction has been quantified in terms of $-\Delta G^\circ$, other questions arise. The answers to 'How is this quantity to be measured and put to practical use?' must be deferred, but the question 'What precise, quantitative meaning is to be attached to "tendency to occur"?' needs attention. This form of words is conventional, and might evoke the irritable remark that either something occurs, or it does not—which is it to be? This might be to the point for chemical reactions on the microscopic scale, but not on the macroscopic or thermo-dynamic scale—'balanced reactions' leading to mixtures of reactants and products coexisting in equilibrium are encountered early in chemical studies. 'Tendency to proceed' would be a better phrase, because 'proceeds to what extent?' is a natural sequel, reasonably expected to elicit a quantitative answer—suited to quantify the 'tendency'. Such an answer is given by the *equilibrium constant*. For example, the extent to which the 'ammonia syn-thesis' reaction proceeds (in presence of a catalyst to abolish kinetic barriers) can be expressed by K_P, an equilibrium constant determined by the partial pressures of the participating gases in the equilibrium mixture:

$$N_2 + 3H_2 \rightleftharpoons 2NH_3$$

$$K_P = \frac{(P^e_{NH_3})^2}{P^e_{N_2}(P^e_{H_2})^3}$$

where P^e stands for equilibrium partial pressure.

At each temperature, K_P is independent of the total pressure* (excepting the influence of gas non-ideality).

Less familiar, perhaps, is the concept that *all* reactions, involving gaseous substances or not, have an equilibrium constant of the nature of K_P. The relation between K_P and $-\Delta G°$ for a reaction is

$$K_P = \exp(-\Delta G°/RT) \tag{2.5}$$

where R is the gas constant per mole, and T is the temperature in $°K$.

Splitting $-\Delta G°$ into its contributory terms, on the lines of equation (2.4),

$$K_P = \{\exp(-\Delta H°/RT)\}\{\exp(\Delta S°/R)\} \tag{2.6}$$

Alternatively,

$$-\Delta G° = RT \ln K_P \tag{2.7}$$

which is the usual form of the relationship known as the *reaction isotherm*, or, now less usually, the van't Hoff Isotherm.

Two essential points must be made. First, $\Delta G°$ is to be regarded in general as $G°_{products} - G°_{reactants}$, i.e. a difference between like extensive properties of products *in their standard states* and reactants *in their standard states*. Chemical reactions do not proceed from initial to final states of this convenient nature—they more often than not proceed from reactants in nonstandard states to a final equilibrium state, sometimes an obvious mixture, characterised by K_P. Hence the often-used definition of $\Delta G°$ as 'the standard Gibbs free energy change *accompanying* the reaction . . .' is misleading and requires translation. As a reminder, the alternative notation

$$\Delta G° = \Sigma n G° \tag{2.8}$$

is useful. In this, n is a stoichiometric number (number of moles of each participant in the reaction) counted positive for products, negative for reactants.

---

* If x is the fractional conversion of nitrogen to ammonia at equilibrium, $K_P = 16(2-x)^2/27P^2(1-x)^4$, where P is the total pressure. At first sight this seems to contradict the statement that K_P is independent of P, but, of course, x depends on P in such a way that K_P remains constant when P is varied.

The second point is that K_P in equations (2.5) to (2.7) must obviously be dimensionless, whereas, except for isomolecular reactions (equal numbers of moles of products and reactants) it is *not*, and must be expressed in appropriate units; e.g. for the ammonia synthesis, as set out above, K_P is a (pressure)$^{-2}$. This often shelved difficulty is met, pending later comment, by noting that, in the forms of the isotherm quoted, K_P is to be divided by the unit used in its measurement.

Although the proof of this important equation comes later, what it is saying can be grasped intuitively, and it is highly desirable that it should be. It is obvious that if the affinity of a reaction is great ($\Delta G°$ large and negative), then, for a given T, K_P is correspondingly large, i.e. the products (always represented in the numerator of K_P) are strongly predominant in the final equilibrium. It is also obvious (but was not before) that even if $\Delta G°$ is positive, K_P does not vanish. The situation is illustrated in Fig. 2.1, which is a graphical representation of equation (2.5) for equilibria at 25 °C; four scales of ordinates (K_P) have been used, each differing from its neighbours by one order of magnitude, against a common scale of $\Delta G°$. It is seen that all the curves are identical in shape, and they are but members of an infinite sequence extending indefinitely in either direction. Where is the line to be drawn between reactions that 'occur' and those that do not? Clearly, this is to be decided in the context of the problem in hand. It is useful to remember that when $\Delta G° = 0$, $K_P = 1$.

An insight into the influence of temperature can be gained by writing equation (2.6) in the form

$$\ln K_P = \frac{-\Delta H°}{RT} + \frac{\Delta S°}{R} \tag{2.9}$$

If, as an approximation, it can be assumed that $\Delta H°$ and $\Delta S°$ do not vary with temperature, it can be seen that the higher the temperature, the greater the proportional part played by $\Delta S°$ in determining a given equilibrium state. This is consistent with previous discussion. If equation (2.9)—on the basis of the same approximate assumption—is differentiated with respect to temperature (all other significant independent variables being kept constant), the result is

$$\partial \ln K_P / \partial T = \Delta H° / RT^2 \tag{2.10}$$

This is known as the van't Hoff equation.* Not at present to be proven, it can be seen to incorporate the familiar le Chatelier principle in relation to the effect of change of temperature on a chemical equilibrium. Thus,

---

* Frequently miscalled for historical reasons, the van't Hoff Isochore, a name suited to the condition of constant volume, not appropriate to equation (2.10).

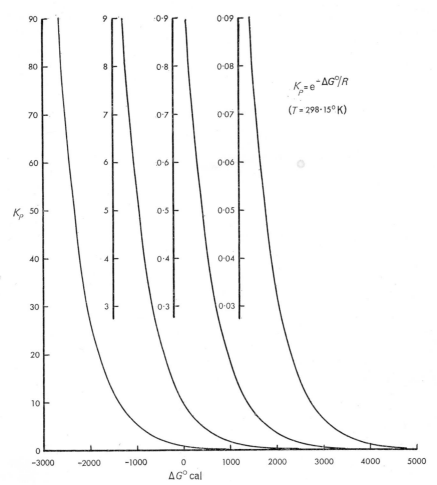

$$K_P = e^{-\Delta G^\circ/R}$$

$$(T = 298 \cdot 15^\circ \, \text{K})$$

Fig. 2.1   Graphical representation of the dependence of K_P on ΔG°

for an endothermic reaction (ΔH° positive), rise of temperature shifts the equilibrium in favour of the products (K_P increases), i.e. the system reacts in a sense to absorb heat. Endothermic reactions are fostered by high temperatures. For exothermic reactions (ΔH° negative), the reverse is the case—for optimum yields of products, high temperatures are to be avoided, and catalysts are often to be looked for.

At this point there is a natural break in the narrative, and an opportunity to recall the objective and assess progress. A step has been taken in showing how to get answers to the first question to be asked about reactions. A few thermodynamic equations have been partially translated. At least a clue has been given as to the interest and powers of the subject in chemistry, and why the effort to study it systematically will pay dividends. Answers must now be sought to the second question to be asked about reactions—the kinetic one—needed for balance in viewing chemical reactivity.

The question can be framed 'What puts the brakes on chemical reactions?'

The first answer is, hindered access of reactants to each other. From recorded standard data, it is found that $\Delta G°$ for the reaction $Mg + \frac{1}{2}O_2 = MgO$ at 25° C and 1 atm is about -140 kcal, so the affinity of the reaction is huge. Yet magnesium is unreactive under these conditions, otherwise it could hardly be used in constructional alloys. Ignited, it burns like fury, suggesting that thermodynamics comes into its own at high, but sometimes not at low, temperatures. The reason is that the metal, in common with most, develops a very thin, adherent and coherent oxide film on its surface—a passivating film that exerts a protective action. If thermodynamics had its way, all except the noble metals would quickly disintegrate into crumbly corrosion products. This is but an extreme example of a kind of *rate-limitation* to be anticipated for any *heterogeneous reaction*, i.e. one which takes place at an interface between dissimilar phases (solid/gas, solid/liquid). Even if there is no inert barrier layer, a reactant must come to the interface, the reaction must proceed, and the product must leave to make room for more reactant. However intrinsically fast the reaction, the over-all reaction rate may be controlled by transport processes (diffusion, convection). The problems concerned are said to be those of *mass transfer*. Important as they are, they need not enter further in the present 'basic' review, in which the writer is again skipping little enough.

The main concern is with *homogeneous reactions*; those occurring within a single gas or solution phase. Since all chemical reaction depends on molecular encounters, reaction rate must depend fundamentally on numbers of molecules of reactants per unit volume, i.e. on concentrations. Reaction rate is normally measured in terms of a rate of change of concentration. For a formalised reaction

$$A + B = C + D$$

Initially a b 0   0   ⎫

At time t $(a-x)$ $(b-x)$ x x ⎬ concentrations, mole l^{-1} ⎭

The rate at time t may be expressed alternatively as

$$dx/dt = dC_C/dt = dC_D/dt = -d(a-x)/dt = -d(b-x)/dt$$
$$= -dC_A/dt = -dC_B/dt \qquad (2.11)$$

where C_C represents the concentration of the product C, and so on.

For a reaction of this stoichiometry, it is usually found (disregarding, now and throughout, non-ideality) that the differential rate equation is

$$dx/dt = k_2(a - x)(b - x) \qquad (2.12)$$

where k_2 is the *velocity constant, rate constant* or *specific reaction rate*. It is seen to have the dimensions* (mole $l^{-1})^{-1} t^{-1}$ or l mole t^{-1}, and to be the reaction rate for the special case of both concentrations maintained at unity. Such a reaction, with rate proportional to the product of two concentration terms (or to the square of one term) is said to be second order—this is the reason for the subscript in k_2. For a generalised reaction, such as

$$a\text{A} + b\text{B} + \ldots n\text{N} = \text{products}$$

where $a, b, \ldots n$ are stoichiometric numbers, the rate law could conceivably be

$$-dC_A/dt = kC_A^a C_B^b \ldots C_N^n \qquad (2.13)$$

If such were the *experimentally determined* rate law the reaction would be said to be of ath order with respect to reactant A, bth order with respect to reactant B, and nth order with respect to N. The '*order of reaction*', however, means the over-all order, and is the sum of all the exponents of the concentration terms in the experimental differential rate equation, i.e. $a + b + \ldots n$. In practice, no order greater than the third is found.

First order reactions, with rate conforming to

$$dx/dt = k(a - x) \qquad (2.14)$$

are of particular interest because it seems odd that there should be such reactions. If reaction depends on molecular encounters, and two molecules are needed for an encounter, a squared concentration term would be expected in the simplest differential rate equation. The present purpose is not to explain this, but to make the point that no direct inference can be drawn from reaction order about reaction mechanism. Reactions often proceed by sequential steps, one of which is rate-controlling. When investigation has shown that the rate-determining step involves an encounter between two molecules, it can then be said that the reaction is *bimolecular*. In general, a statement about the *molecularity* of a reaction refers to the mechanism by which it proceeds.

* Not strictly a proper use of the word, in the sense that the quantity is shown as a function of contributory, but not 'basic' quantities.

With these minimal guide-lines established, attention can be given to kinetic aspects of chemical equilibrium in terms of the formalised 'reversible reaction'

$$A + B \rightleftharpoons C + D$$

From whichever side it is approached, identically the same dynamic equilibrium state is attained; then, continuing forward and back reactions have zero net effect because they are proceeding at equal rates. There is a normally imperceptible *exchange process* essential to preservation of equilibrium. If, by reason of quenching from a higher to a lower temperature, the *exchange rate* should tend to zero, the higher temperature state is 'frozen in', and becomes a non-equilibrium state at the lower temperature.

The fact that dynamic equilibrium exists can be expressed

rate of forward reaction $= k_2 C_A C_B$ — equal at equilibrium
rate of back reaction $\quad = k_{-2} C_C C_D$

therefore

$$k_2/k_{-2} = C_C C_D / C_A C_B = K_C \qquad (2.15)$$

where K_C, seen to be the equilibrium constant in terms of concentrations, is equal to the ratio of the rate constants for the forward (k_2) and back (k_{-2}) reactions. Clearly, a given value of K_C can arise in an infinity of ways—k_2 and k_{-2} can vary between small and great, as long as their ratio remains the same.

The influence of temperature must now be considered, but this has already been done in relation to the equilibrium constant. Equations (2.5) to (2.10) apply to K_C as to K_P, the only essential change being redefinition of standard states. It is, however, appropriate to rewrite equations (2.9) and (2.10) suitably to the present context, i.e.

$$\ln K_C = - \frac{\Delta H^\circ}{RT} + \frac{\Delta S^\circ}{R} \qquad (2.16)$$

$$\partial \ln K_C / \partial T = \Delta H^\circ / RT^2 \qquad (2.17)$$

It remains evident that $\ln K_C$ (and K_C) has a positive or negative temperature coefficient according to the sign of ΔH°—whether the system, in reacting, climbs, or descends, an energy step.

The situation in relation to rate constants is different. They all have remarkably high, positive temperature coefficients, typified by more than doubling of reaction rate for a 10° rise in temperature. This suggests that reacting systems have to climb a comparatively precipitous energy step on their way to the final goal of *having formed products*, i.e. there is indeed a barrier to reaction.

The dependence of rate constant on temperature is normally well expressed by the equations

$$\ln k = - \frac{E_a}{RT} + \text{constant} \tag{2.18}$$

$$\partial \ln k / \partial T = E_a / RT^2 \tag{2.19}$$

where, for each reaction, E_a is a constant, positive quantity with the dimensions of energy. That these are similar in form to equations (2.16) and (2.17) should not occasion surprise—it could hardly be otherwise in the light of equation (2.15).

Consider again the balanced reaction (bearing in mind that, in principle, all reactions are balanced). If, for the forward reaction,

$$\partial \ln k_2 / \partial T = E_{a(2)} / RT^2$$

and, for the back reaction,

$$\partial \ln k_{-2} / \partial T = E_{a(-2)} / RT^2$$

then, since $K_C = k_2 / k_{-2}$, or $\ln K_C = \ln k_2 - \ln k_{-2}$, it is evident that

$$\Delta H^\circ = E_{a(2)} - E_{a(-2)}$$

ΔH° now appears as a difference between two energy terms, both positive. At least one of these terms must be as large in magnitude as ΔH°, and the facts indicate that both are larger. This situation is represented in Fig. 2.2 (drawn appropriately to an exothermic reaction). The axis of abscissae depicts the, at present, not very clearly definable 'reaction co-ordinate', increase of which marks the progress of reaction. $E_{a(2)}$ and $E_{a(-2)}$ are called the *activation energies* of the forward and back reactions.

It is the task of theories of reaction kinetics to interpret these basic facts. For a bimolecular reaction in the gas phase—the simplest to consider—it could be thought that reaction might occur whenever molecules of the reactants happened to collide in the course of their random thermal motions. But the kinetic theory of gases gives information (in terms of molecular dimensions, numbers of molecules in unit volume and temperature) which excludes this idea—it would require very great reaction rates increasing linearly with $T^{\frac{1}{2}}$. A suggestion by Arrhenius (1889) overcame this difficulty, and was basic to the development of a tenable *collision theory* of chemical reaction. It was that there is in every system an equilibrium between 'normal' and 'active' molecules, and that only the latter are capable of reaction. This

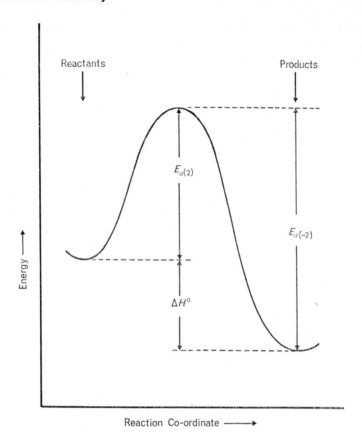

Fig. 2.2   Representation of reaction energy profile

remains valid in the sense that there is in every system an equilibrium distribution of molecular energies, and it is only those collisions to which the participating molecules contribute sufficient energy that are effective in bringing about reaction. Renewed appeal to the kinetic theory of gases elicits the fact that at temperature T, the proportion of gas molecules with kinetic energy not less than a critical value E is simply

$$\exp(-E/RT)$$

If it is indeed only these molecules that can react, a theoretical rate law for the simple gas reaction might be

$$k_2 = Z \exp(-E/RT) \tag{2.20}$$

where Z is the *collision number* calculable from kinetic theory, i.e. a collision frequency per unit volume so standardised in relation to concentration as to make rate equal to specific reaction rate. The experimental rate law of equation (2.18) can be written

$$k_2 = A \exp(-E_a/RT) \qquad (2.21)$$

where the constant, A, is called the *frequency factor*, or the *pre-exponential* factor. The question at once arises, whether A can be identified with Z for a suitably simple gas reaction, disregarding in the first place the comparatively small dependence of Z on temperature. It turns out that this identification can be made in a few cases (e.g. $2HI = H_2 + I_2$), but mostly *not*. For reactions at large (excluding chain reactions with their obvious consequences), A is smaller than Z, often by several orders of magnitude. This points to the existence of yet another factor fundamental to reaction rate, such as to permit only a small proportion of even sufficiently energetic collisions to be fruitful of reaction. It is easy to see that the 'hard sphere' (billiard ball) model of the simplest collision theory is not suited to molecules with structures and specific reaction sites. Very special steric and orientational requirements of the two molecules may need to be satisfied at the moment of encounter for reaction to take place. For this reason, a 'catch-all coefficient', known as the *steric factor* or *probability factor*, P, is introduced into equation (2.19). Generalised for a reaction of any order, this equation reads

$$k = PZ \exp(-E_a/RT) \qquad (2.22)$$

and is commonly known as the Arrhenius equation.

The *transition state theory* is an alternative approach of wider scope, bringing in *rate processes* other than chemical reactions, such as diffusion, or the flow of liquids. The view this theory takes may be described as 'gradual', envisaging, for example, bond making and breaking in chemical reactions as overlapping and co-operative processes. It can be introduced by consideration of interatomic distances. The distances between atoms covalently bound together (internuclear distances, bond lengths) are short and accurately measurable. The C—C bond, for example, is 1.54 Å long, and varies little in length from one saturated aliphatic compound to another. On the other hand, the minimum distance between non-bonded atoms is considerably greater—strong repulsion comes into play when atoms belonging to different, non-reacting molecules approach closer than about 3.5 Å. Between bonded and non-bonded interatomic distances, there is a 'forbidden range' of lengths very seldom found in stable systems. Yet, in the course of a reaction involving the breaking and making of bonds, there must be a transition state in which these abnormal interatomic distances exist. Between reactants and products, there must be transitory entities containing abnormally elongated, half-broken bonds, and abnormally long bonds-in-the-making, which must be intrinsically unstable and of high energy content. This is a way in which an *activated complex*, or *transition state* can be pictured.

The energy plot of Fig. 2.2 is still applicable, the energy peak now corresponding to the transition state, which is precisely what its name implies—it is always being passed through by entities about to roll down the slopes on either side, to form products, or re-form reactants. It must not be confused with an intermediate compound, because it lies at a maximum, not a minimum of free energy. Nevertheless, in the development of the theory, extensive thermodynamic properties are assigned, questionably, to the activated complexes or 'transition states'. Justification is found in the results.

If it is assumed that, during reaction, equilibrium is maintained between reactants and transition states, e.g.

$$A + B \rightleftharpoons AB^{\neq} \rightarrow \text{products}$$

with an equilibrium constant

$$K_C^{\neq} = C_{AB^{\neq}}/C_A C_B \tag{2.23}$$

the theory leads to a rate law

$$k_2 = kTK^{\neq}/h \tag{2.24}$$

where k is Boltzmann's constant, $1 \cdot 3805 \times 10^{-16}$ erg $^{\circ}K^{-1}$/molecule, and h is Planck's constant, $6 \cdot 6256 \times 10^{-27}$ erg s. It is seen that kT/h has the dimensions t^{-1}, and acquires the significance of the rather low vibration frequency of the abnormal, weak bond in the activated complex; the vibration that leads to its flying apart to form products.

If, now, the reaction isotherm (equation (2.7)) is applied in the form

$$-\Delta G^{\neq} = RT \ln K^{\neq} \tag{2.25}$$

where $\Delta G^{\neq}$ is a standard Gibbs free energy of activation, the rate law becomes

$$k_2 = \frac{kT}{h} \exp(-\Delta G/RT) \tag{2.26}$$

and, using the appropriate analogue of equation (2.4), i.e.

$$\Delta G^{\neq} = \Delta H^{\neq} - T\Delta S^{\neq} \tag{2.27}$$

the rate law is expanded to

$$k_2 = \frac{kT}{h} \{\exp(\Delta S^{\neq}/R)\}\{\exp(-\Delta H^{\neq}/RT)\} \tag{2.28}$$

which bears immediate comparison with the experimental rate law of equation (2.21), namely,

$$k_2 = A \exp(-E_a/RT) \tag{2.21}$$

The pre-exponential factor, A, takes on a new significance as a function of an *entropy of activation*, $\Delta S^{\neq}$, whilst the activation energy E_a, remains determined by (but not precisely equal to) the enthalpy of activation, $\Delta H^{\neq}$.

Is it not clear that, if there are some special steric or orientational conditions necessary for the formation of the activated complex, and therefore for the consummation of reaction, this will call for the establishment of a very specific kind of order from an initially disordered system of reactants? If so, $\Delta S^{\neq}$ can be expected to be negative, to an extent increasing with the complexity of the necessary organisation in the transition state. It can then be said that the more negative is the entropy of activation, the less probable each elementary act of reaction will be, and the less will be reaction rate.

This importation of thermodynamics into reaction kinetics carries immense dividends. We have in $\Delta G^{\neq}$ a function which is all-inclusive in determining specific reaction rate, just as ΔG° determines equilibrium. Wishing to look deeper, we can determine and examine, in either case, the enthalpy and entropy terms that contribute to this simplifying, 'portmanteau function'. Indeed, all the thermodynamic quantities become of interest in kinetic studies, including heat capacity of activation ($\Delta C_P^{\neq}$) and volume of activation ($\Delta V^{\neq}$); each, to be reconciled with realistic models, represents a deeper digging towards fundamental understanding of chemical reactions—they are the basis of very active current research.

It would be unfortunate to leave an impression that the collision theory is outdated and moribund. This is not so; it also has its sophisticated, modern versions, as witness a recent Faraday Society discussion. The complementary nature of views from different angles is thus illustrated, and may perhaps be reinforced in a penultimate paragraph.

There may be several 'chemical equilibrium states' in which a sufficiently versatile chemical system may exist. They can be regarded as free energy valleys of varying elevation, falling under the purview of thermodynamics. They are separated from each other by ranges of high ground—saddlebacks and peaks of varying altitude, corresponding to non-equilibrium states. The high ground, of course, is necessary for the clear definition of the valleys. How traverse from one valley to another can be accomplished is a matter for reaction kinetics, and the journey may be relatively easy (balanced reactions) or almost impossibly difficult. In the latter case, reserves of energy must be summoned up (by raising the temperature), or other means found for reducing the effective height of the barriers (by looking for a catalyst).

Now although it is the essence of a catalyst that it has no effect upon the balance of 'an equilibrium', it is also of its nature to be highly specific. A given catalyst will effectively truncate *one* kind of peak, another catalyst will partially abolish another kind of barrier, and so on. This is why alumina at 400 °C facilitates the 'dehydration' of ethanol to form ethylene and water, but nickel at 400 °C promotes its dehydrogenation, forming acetaldehyde and hydrogen—a standing puzzle to students because of its incompatibility with over-simplified concepts of catalysis, and also because of too narrow a view of the meaning of 'equilibrium', not encompassing its relative nature.

The review of 'chemical reactivity' is now concluded, and the reader must judge its relevance and degree of success in demonstrating the indispensability of the thermodynamic approach—*coupled with others.*

3

Thermochemistry

3.1   Introduction

This long chapter requires an explanation. It starts on traditional lines, but soon digresses to discuss heat capacity at some length. This is because the subject is interesting and well illustrates how profitable it is to think about the significance of thermodynamic functions. After this, normal development is resumed, but it leads into an extensive essay on the application of thermochemistry in chemistry at large. This is again because it is interesting, and important in modern inorganic chemistry.

Strictly, thermochemistry deals only with heat transfers involved in chemical reactions occurring irreversibly—as in calorimeters. So restricted, it is but a part of thermodynamics, disciplined only by the first law.

The foundations of thermochemistry were laid in the last century by Berthelot and Thomsen, who found impetus for their extensive work in the belief that the affinity of a chemical reaction was to be measured in terms of heat evolution. Although mistaken in this (as they later realised under the guidance of Nernst), their monumental contribution stands—the establishment of 'heats of formation of compounds' as the first chemical-thermodynamic function to be systematically determined and critically tabulated.

Thermochemistry is dominated by the experimental science of calorimetry, with a literature so large, and techniques so varied and specialised, as to place it beyond the scope of this book. The results must be accepted with respect—and also with discrimination, because some of the most potentially rewarding calorimetric problems are of the greatest experimental difficulty; calorimetric research actively continues.

Calorimetry itself was instrumental in removing the original restriction on thermochemistry by opening the way to evaluation of absolute entropies of pure substances—a matter to be considered in due course. This made available *one* route to the valid assessment of affinities in terms of $\Delta G° = \Delta H° - T\Delta S°$. This is perhaps why the name 'thermochemistry' is occasionally used to embrace the whole of chemical thermodynamics.*

It is somewhat unfortunate that the label 'the erroneous Thomsen–Berthelot principle' should have been attached to the conclusions of two great pioneers. The repellent title is destructive of a sense of proportion. Accepting the error, are there perhaps circumstances in which the principle is a useful yardstick, or in which the error can be reduced to insignificance? We know that it is $-\Delta G°$, and not $-\Delta H°$, as supposed by Thomsen and Berthelot, that is the proper measure of affinity of 'macroscopic processes'. Yet, bearing in mind that many chemical reactions involve large energy changes (bond making and breaking), $-\Delta H°$ may not be a bad assessment of $-\Delta G°$, provided that the $T\Delta S°$ term is relatively small. This condition is likely to be reasonably satisfied for reactions with all product and reactants in solid or liquid states, to the exclusion of gases or solutes. Another such circumstance is that the ultimate objective is to deduce from the behaviour of macroscopic systems information about microscopic events, such as the formation of single chemical bonds; $-\Delta H°$ may then be sounder as a basis of reasoning than $-\Delta G°$. Yet another is to seek the condition under which these two functions become identical. How these alternative circumstances arise is to be illustrated in due course.

Before proceeding further, there is a difficulty to be ventilated, important enough to require its own sub-heading.

3.2   A language difficulty

There is an unresolved, widespread source of confusion in thermochemical discussion—not the only one of its kind.

Partly because many more chemical reactions evolve heat than absorb it, the founding thermochemists, and succeeding generations, counted the heat evolved in exothermic reactions as a positive quantity. Since the classical work was carried out in this continent, this became the European convention. In the older notation (still not infrequently used), heat produced was represented, along with other products, in thermochemical equations of the following type

$$H_2(g) + \tfrac{1}{2}O_2(g) = H_2O(l) + 68\cdot32\,\text{kcal}$$

or, counting heat absorbed in endothermic reactions as negative, by, for example,

$$\tfrac{1}{2}H_2(g) + \tfrac{1}{2}I_2(s) = HI(g) - 6\cdot20\,\text{kcal}$$

* Cf. *Metallurgical Thermochemistry* by O. Kubaschewski and E. Ll. Evans, Pergamon Press, 3rd ed., 1958, which contains a survey of calorimetric methods.

In these two examples, the 'heats of reaction' would be 68·32 kcal and −6·20 kcal respectively. If, further, the tacit implication of standardisation of state be accepted (25 °C, 1 atm), these quantities would also be the standard 'heats of formation' of liquid water and gaseous hydrogen iodide.

The first, minor, difficulty comes in fitting this older, but reasonable and self-consistent, scheme into the wider framework of modern thermodynamics. It is necessary to identify positive 'heat of reaction' with enthalpy loss, $-\Delta H$; this requires a slight effort of memory. Some authorities steadfastly maintain this nomenclature and translation. Others do not. This is because, over the years, the centre of gravity of thermochemical work moved across the Atlantic, and an American convention grew up, requiring 'heat of reaction' to be re-defined as heat *absorbed* during the reaction, and therefore identical with enthalpy gain, ΔH. Setting this out:

Europe:

heat of reaction = heat evolved = $-\Delta H$

America:

heat of reaction = heat absorbed = ΔH

The effort of memory required is transferred from the second equality to the first. Perhaps it is time to abolish old, 'pre-thermodynamic' conventions, but the major difficulty lies in reversing so long-standing and traditional a usage. Nevertheless, this is the trend, and students must be alive to it.

The situation is this. It is common parlance, from which chemists are unlikely to be shifted, to speak, and write, of 'heat of . . .' in relation to combustion, neutralisation, fusion, evaporation, solution, dilution, ionisation, adsorption and many another process. Some of these are always exothermic, some are always endothermic, others may be either the one or the other from case to case. This colloquial usage will undoubtedly continue, and is unobjectionable when no ambiguity can arise. It is certainly impracticable to get everyone to *mean* 'ΔH of . . .' when they *say* 'heat of . . .'—or to quote the calorific value of a fuel in terms of a negative heat absorption per kg burned. Of course, if we could all school ourselves to use only 'enthalpy of . . .', clearly meaning ΔH, we should be compelled to attach a correct sign to any value in kcal we gave, and all ambiguity would vanish. Fortunately, this is precisely what *has* been universally agreed in setting out thermochemical equations, as, for example,

$$H_2(g) + \tfrac{1}{2}O_2(g) = H_2O(l); \; \Delta H°_{298·15} = -68·32 \text{ kcal}*$$

where the superscript to ΔH indicates that reactants and products in the molar units of amount indicated in the chemical equation, and in the states

* A reminder: *not* kcal mole^{-1} for a stated *reaction*.

of aggregation shown by symbols in brackets, are in standard states determined by fixing pressure and temperature at standard values (P, tacitly, 1 atm; T specified in °K by the subscript to $\Delta H°$).

There is, therefore, a problem still for readers of the literature, but, with forewarning, not a difficult one—except when they meet a definition of 'heat of formation' which runs 'the quantity of heat *liberated or absorbed* when, etc., etc.', which muddles up the whole issue.

3.3 Enthalpy and its temperature dependence at constant pressure

To recapitulate, enthalpy, H is an extensive property of a system, not determinable on an absolute scale. The change in enthalpy, ΔH, of a system caused by any process *is*, however, determinable, and is strictly the difference between the enthalpy of the system in its final state after completion of the process (H_2) and the enthalpy of the system in its initial state, before the process occurred (H_1). Regarding enthalpy as total energy, this is consistent with the first law—the law of conservation of energy. Hence, $\Delta H = H_2 - H_1$. If the process occurs irreversibly in an open calorimeter at constant pressure,

$$\Delta H = q_P \tag{3.1}$$

where q_P stands for heat absorbed at constant pressure.

The simplest process to consider is perhaps that resulting from a change of temperature alone—a change of but one of the independent variables of state, the other, pressure, being kept constant. This raises the general problem of the rate of change of enthalpy with respect to temperature, which can be discussed rigorously only in terms of the appropriate partial differential coefficient,* namely,

$$C_P = \left(\frac{\partial H}{\partial T}\right)_P \tag{3.2}$$

which defines C_P, the heat capacity at constant pressure. Since H is an extensive property of a system, C_P must be, too. If the system under consideration happens to be one mole of a pure substance, C_P is a molar heat capacity at constant pressure, but this is not always the appropriate translation, and care is necessary.

If the enthalpy of a system under constant pressure is plotted against temperature, the resulting line can only be placed arbitrarily against the vertical axis because there is no agreed zero for an enthalpy scale. For the

---

* It may enter the reader's mind that it is 'specific heat' that is to be discussed. There are three reasons, which the reader is invited to spot for himself, why it is not.

simplest of systems, such as a mole of a monatomic gas, the plot is rectilinear over a wide temperature range, but for most systems it is not (Fig. 3.1a). This is why it is obligatory to define heat capacity in terms of a partial differential coefficient, the graphical translation being the slope of the enthalpy-temperature plot at any precisely stated temperature.

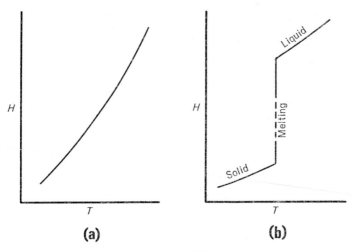

Fig. 3.1 Schematic enthalpy/temperature plots
  a. For a system undergoing no transition
  b. For a solid melting within the temperature range concerned

It appears, then, that for most systems, heat capacity is itself a function of temperature. Why should this be so? Answers to this question in specific cases have led to important theoretical advances, but it would be premature to attempt discussion of them at the present stage. But there is good reason to think about a generalised answer—indeed the writer considers it essential for students to do so in the interests of translating sterile-appearing thermodynamic functions into discussible physical realities. The following paragraphs are illustrative.

Consider an experiment to measure the heat capacity at constant pressure of a pure solid substance over a range of increasing temperatures. This can be done by metering in small quantities of energy by electrical means, and measuring the resulting small increments in temperature; the method can be quite accurate, and can give results not differing significantly from the required partial differential coefficient. The results provide a smooth plot until the melting point of the solid is reached. It is then found that large quantities of energy can be pumped into the system with no consequent rise of temperature. This behaviour persists until all the solid has melted, when rise of temperature is resumed for each further intake of energy. What happens is formally illustrated in Fig. 3.1b—the plot of enthalpy against temperature at constant pressure contains a precisely vertical section. So, whilst melting is going on, it appears that C_P, as defined by equation (3.2), is infinite.

This raises a debating point. It could be maintained that the enthalpy increase at constant temperature is a latent heat of fusion, and that it is not sensible to interpret it in terms of an infinite heat capacity of the two-phase, solid–liquid system; instead, C_P must be regarded as a discontinuous function of temperature. Certainly, all changes of state which take place when temperature is raised are accompanied by absorption of latent heat, but the expression 'change of state' is usually used in the sense of a change of 'state of aggregation' (solid → liquid; liquid → vapour; solid phase transition). This is an arbitrary restriction on the meaning of the word 'state', and there is a great deal more to think about.

It is useful to enlarge on the earlier theme of energy-entropy balance, and to reflect that crystalline solids exist because 'structural forces' are able to hold the basic building units together in ordered, three-dimensional array against the disruptive influence of thermal motions. With rising temperature, these motions become more and more energetic until, at the melting point, they prevail, and long-range order is destroyed. Still, however, the forces of attraction succeed in holding the molecules, atoms or ions together, and cohesion is retained in the *disarray* of the liquid state. At some higher temperature, cohesion is overcome, and the liquid passes into disordered vapour. Melting and evaporation, occurring at unique temperatures, are unmistakable defeats of order-producing forces under the attack of disruptive thermal motions. They are *first-order transitions*—marked by continuity in G (phases in equilibrium at the transition temperature) and discontinuity in S (entropy jumps). The conflict, however, goes on all the time, and the balance shifts towards increasing entropy by any available means as the temperature rises. Every little weakness is exploited, and disruption may secure minor advances distinct from decisive victories, in, for example, the following way.

Some solids undergo *second-order transitions* (with continuity of both G and S), arising from order-disorder transformations taking place over a temperature *range*, without structural collapse. Many alloys show this behaviour, but it is best to use a pure substance in illustration, namely, solid methane. At very low temperatures, the tetrahedral, non-rotating CH_4 molecules have orderly, correlated orientations in the crystal lattice. With rise of temperature, however, molecular rotation can set in (with increase of entropy) without damage to the crystal structure, because of the high degree of symmetry of the molecules. It is interesting to consider the likely course of events, bearing in mind the distribution of thermal energy in an assembly. At first, a few molecules start rotating, losing specific orientation, but this is no great matter. At a slightly higher temperature, many more may have become orientationally disordered, but still a 'majority vote' would leave no doubt about the distinction between 'correct' and 'incorrect' orientations. Very soon, however, serious doubt will arise; each molecule, instead of being fortified in rectitude by the majority of its neighbours, will be infected by lawlessness. Once the rot has become extensive, complete collapse of order cannot be long delayed. This is indeed the way in which *co-operative order* fails,[1] and it is illustrated in Fig. 3.2a, where it is seen that loss of order is complete at a specifiable *lambda point*, T_λ. The Curie point—the temperature above which a ferromagnetic metal cannot retain permanent magnetism—is a lambda point.

Second-order transformations give rise to heat capacity anomalies. These are peaks in the $C_P(T)$ curves, variable in height, width and shape from case to case (but crudely represented by a Greek capital lambda; very steep on

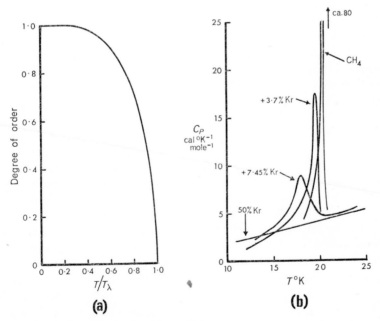

(a)   **(b)**

Fig. 3.2 (a) Order–disorder transformation
(b) Heat capacity 'anomalies' in solid methane–krypton systems

the high temperature side). A very interesting result came from the study of methane progressively diluted (in solid solution) with krypton;[2] this is illustrated in Fig. 3.2b, and shows the intelligible effects of reducing the restraints between neighbouring methane molecules.

It would be out of context to continue this discussion, except to draw attention to the normal, steady changes that, in absence of eccentricities, must accompany rise of temperature. Is it not clear that a solid, for example, must suffer a change of state, when its temperature is raised, because of thermal expansion? The mean intermolecular distance is increased, and the close-range, distance-sensitive, intermolecular forces are weakened. It is hardly surprising that the heat capacity changes, and it may be thought that C_P at a particular temperature is a function not only of the state of the system concerned at that temperature, but also of the rate at which that state is changing with temperature.

All in all, it becomes not so easy to isolate a 'latent heat' and say when a contribution of this nature does, or does not form part of C_P. Of course, the heat capacities of simple gases and 'monatomic' solids are well understood—they are discussed under the heading of 'experimental foundations of the quantum theory',[3] but this cannot be said of systems in general. Heat capacity, and changes of heat capacity accompanying reactions (or the formation of transition states), remain functions of great physico-chemical interest.

Returning from the digression to more basic matters, it is clear from equation (3.2) that the increase in enthalpy of a system (undergoing no transitions) due to a finite increase of temperature at constant pressure must be found by an integration:

$$\Delta H = H_{T_2} - H_{T_1} = \int_{T_1}^{T_2} C_P . dT \qquad (3.3)*$$

For this, of course, C_P must be known as a function of temperature from experimental determinations extending over the range concerned, and its graphical equivalent is the determination of the area under a $C_P(T)$ plot between the appropriate temperatures. In general, heat capacity studies involve different calorimetric methods and objectives for ranges of temperature below and above room temperature; it is only the latter which are of present concern, as the day-to-day business of practical thermochemistry.

---

* With some compunction, it is remarked that here ΔH refers to $T_1 \to T_2$ at const. P; previously, ΔH has referred to reactants $\to$ products at const. T,P. Emphatically, 'ΔH' means nothing unless referred to a specified change of state of a system. Failure to remember this leads to muddles.

In this connection, the experimental $C_P(T)$ data for pure substances (solid, liquid, gas) can be represented by empirical interpolation equations, e.g. polynomials in absolute temperature,

$$C_P^\circ = \alpha + \beta T + \gamma T^2 \dots \text{cal mole}^{-1} \qquad (3.4)^*$$

Where $\alpha, \beta, \gamma \dots$ are numerical coefficients chosen to give the best fit to the experimental data; they are *parameters*† and have no theoretical significance. With a sufficient number of terms, such an equation is capable of faithful expression of a given set of data with no accuracy (nor inaccuracy) lost or gained (except by legitimate smoothing). How many terms it is sensible to include depends on the intrinsic accuracy of the measurements, the form of $C_P(T)$ (thermal anomalies cannot be accommodated), the range of temperatures, and the purpose of the job in hand. In some cases it may be an acceptable assumption (or an enforced one in absence of data) that $C_P^\circ = \alpha = \text{constant}$. In others, $C_P = \alpha + \beta T$ is adequate, or necessary, pending the provision of extended measurements. It is seldom that more than three terms are needed, even for a $2000°$ temperature range.

It is clearly a matter of principle that such empirical equations should not be used beyond the temperature range of the measurements that gave rise to them, yet the temptation to *extrapolate* is strong. A concession is made allowing cautious extrapolation to higher, but not to lower, temperatures, provided that the classical polynomial equation (3.4) is replaced by the alternative

$$C_P^\circ = a + bT + cT^{-2} \qquad (3.5)$$

where a, b and c are equally empirical coefficients giving the best fit. The reason for this is that the term CT^{-2}, coping with deviation from linearity, has, in contrast to γT^2, a progressively lesser significance with increasing T. This, in general, better suits the experimental facts, and minimises the dangers of extrapolation. Equation (3.5) also has some advantage in easing computations that normally follow; it is therefore replacing equation (3.4) as the generally used form. It is obviously of advantage to standardise the form of such equations because they can then readily be used in conjunction with each other, as shortly to be shown.

When information on $C_P(T)$ for any system or *thing* is available, the

---

* It is not considered necessary to insert units in the formulations that follow.

† This word is increasingly misused. A parameter is a quantity to which the operator may assign arbitrary values, as distinct from a variable, which can assume only those values that the form of the function makes possible. With due adjustment of $\alpha, \beta, \gamma\dots$, or of a, b, $c\dots$, equations (3.4) and (3.5) can fit the same set of data equally well.

integration of equation (3.3) can be carried out; thus if the system is a mole of a pure substance,

$$\Delta H = H_{T_2} - H_{T_1} = \int_{T_1}^{T_2} C_P dT$$

$$= \int_{T_1}^{T_2} (\alpha + \beta T + \gamma T^2 \ldots)\, dT = \left[\alpha T + \tfrac{1}{2}\beta T^2 + \tfrac{1}{3}\gamma T^3 \ldots \right]_{T_1}^{T_2}$$

$$\text{or} \quad \int_{T_1}^{T_2} (a + bT + cT^{-2})\, dT = \left[aT + \tfrac{1}{2}bT^2 - cT^{-1} \right]_{T_1}^{T_2} \tag{3.6}$$

3.4   Hess's Law

This classical law of thermochemistry (1840), sometimes called the law of constant heat summation, or the law of thermoneutrality, is to the effect that the 'heat change' in a chemical reaction is the same whether it takes place in one or in several stages. In thermodynamic notation,

$$\Delta H = H_2 - H_1 \tag{3.7}$$

where H_2 relates to products, and H_1 to reactants, is an adequate statement of the law. ΔH is uniquely determined by final and initial states, and is therefore independent of the path between them. No further 'proof' by appeal to the first law is needed or desirable—the definition, and appreciation, of H as an extensive property is enough.

The importance of Hess's law lies in the establishment of an interlocking system of correlated enthalpy changes, ultimately including all chemical reactions, and, indeed, all processes involving enthalpy change. Each new result plugs into the system, cross-checking and enlarging the body of existing data. There are, of course, corresponding (but unnamed) laws for the other extensive properties, such as Gibbs free energy and entropy, and these also enter into the correlation. It is not difficult to see how fundamental this is to thermodynamics as a quantitative science, and therefore how necessary it is for the student to ponder on, and grasp, the concept of extensive property, on which it all depends.

The first step in this great enterprise was, of course, the accumulation of what we must now call standard enthalpies of formation of compounds, still proceeding with critical selection and tabulation, notably at the National Bureau of Standards, Washington, D.C.[4] These are $\Delta H°$ values (sometimes denoted ΔH_f°) for the reactions in which *one mole* of each compound in its normal state at 25 °C and 1 atm pressure is formed from its constituent elements in their normal, most stable, states* at 25 °C and 1 atm pressure—with the modifying proviso that gases are adjusted to the perfect gas (p.g.) state. This provides, in effect, a scale of standard enthalpies of compounds by arbitrary assignment of zero enthalpy to all the elements in their standard (but no other) states.

Two special examples are

$$C(graphite) + O_2(p.g., 1 \text{ atm}) = CO_2(p.g., 1 \text{ atm});$$
$$\Delta H^{\circ}_{298.15} = -94{\cdot}0518 \text{ kcal}$$

$$H_2(p.g., 1 \text{ atm}) + \tfrac{1}{2}O_2(p.g., 1 \text{ atm}) = H_2O(l);$$
$$\Delta H^{\circ}_{298.15} = -68{\cdot}3174 \text{ kcal}$$

These are formation reactions that proceed quantitatively and can be studied directly; this is one reason why the enthalpies of formation of carbon dioxide and water are known with high accuracy—the other is that they are needed in so many 'Hess's law calculations'. Such calculations are involved in the determination of the enthalpies of formation of most compounds, for it is few that can be formed directly and quantitatively from their elements. On the other hand, many compounds can be burned, with adequately accurate determination of heats of combustion. For instance, the combustion of ethane gives (omitting all but essential indications of state)

A. $C_2H_6 + \tfrac{7}{2}O_2 = 2CO_2 + 3H_2O(l); \ \Delta H^{\circ}_{298.15} = -372{\cdot}87 \text{ kcal}$

But we already have

B. $C(graphite) + O_2 = CO_2; \ \Delta H^{\circ}_{298.15} = -94{\cdot}05 \text{ kcal}$

C. $H_2 + \tfrac{1}{2}O_2 = H_2O(l); \ \Delta H^{\circ}_{298.15} = -68{\cdot}32 \text{ kcal}$

Then by an appropriate algebraic summation of the reactions and of the relevant ΔH° values, namely,

$$2B + 3C - A = D$$

we get

D. $2C(graphite) + 3H_2 = C_2H_6; \ \Delta H^{\circ}_{298.15} = -20{\cdot}19 \text{ kcal}$

which is the desired result. Any reaction lending itself to accurate calorimetry (or to any other method of determining ΔH°) can be included in a 'Hess's law sequence' of reactions, chosen so that each 'unwanted' intermediate

---

* Except for phosphorus; white phosphorus is adopted as standard.

appears *twice*—once as product and once as reactant—and therefore cancels in the final summation. Any intermediate can therefore be in a non-standard state, as long as this state is identically the same for the two reactions in which the intermediate is involved. Thus,

A. $U(c) + 4HCl(aq, 6M) = UCl_4(soln) + 2H_2(g);$

$$\Delta H^\circ_{25^\circ} = -146 \cdot 9 \text{ kcal}$$

B. $UCl_4(c) + HCl(aq, 6M) = UCl_4(soln); \Delta H^\circ_{25^\circ} = -43 \cdot 1 \text{ kcal}$

C. $\frac{1}{2}H_2(g) + \frac{1}{2}Cl_2(g) + aq = HCl(aq, 6M); \Delta H^\circ_{25^\circ} = -36 \cdot 9 \text{ kcal}$

D. $D = A + 4C - B; U(c) + 2Cl_2(g) = UCl_4(c);$

$$\Delta H^\circ_{25^\circ} = -251 \cdot 4 \text{ kcal*}$$

The final result is the standard enthalpy of formation of crystalline uranium tetrachloride.[5]

There is, of course, a two-way traffic between the standard 'formation data' and other reactions; but one example must suffice.

Suppose that one were asked about the following somewhat unprofitable-looking 'dry-way reaction'

$$MgO(s) + Cl_2(g) = MgCl_2(s) + \tfrac{1}{2}O_2(g)$$

Looking up the standard data for an approximate assessment, one would find

A. $Mg(s) + \frac{1}{2}O_2(g) = MgO(s); \Delta H^\circ = -143 \text{ kcal}$

B. $Mg(s) + Cl_2(g) = MgCl_2(s); \Delta H^\circ = -153 \text{ kcal}$

Then $B - A$ gives the desired answer; ΔH° for the proposed 'chlorination reaction' is only -10 kcal. Since (as one might spot) there is also, almost certainly, a decrease in entropy (net loss of $\frac{1}{2}$ mole of high-entropy gas) the prospects are unhopeful. However, the bright idea might occur—why not add some carbon, and look instead at a possible reaction

$$MgO(s) + Cl_2(g) + \tfrac{1}{2}C(graphite) = MgCl_2(s) + \tfrac{1}{2}CO_2(g)$$

This would make the reaction more exothermic (ΔH° more negative) to the tune of 47 kcal, with a similar entropy disability. Indeed, the idea would have been sound.

The reader will have already realised that this example is premature to the point of absurdity on two counts. Why not go directly to the tabulated standard Gibbs free energies of formation of compounds and reach a more

---

*As might be supposed, there is considerable variation in notation; (c) indicates crystalline and aq stands for solvent water—'aqueous'.

decisive conclusion in the first place? Further, what is the use of messing about with such reactions at the standard temperature of 25 °C, when for any practical utility a temperature range up to 2000 °C or more must be considered? The writer, unoffended, replies that this is what he is working up to, but there is still quite a long way to go. The situation is that 25 °C thermochemistry is the essential starting point (for some problems it is also enough), but suffers from disabilities to be successively removed. The first of these is restriction to a single temperature.

3.5   The Kirchhoff theorem

This is the second classical law of thermochemistry (1858), and may be derived from the first, in terms of equation (3.7), namely,

$$\Delta H = H_2 - H_1$$

by differentiation with respect to temperature at constant pressure, thus:

$$\left(\frac{\partial \Delta H}{\partial T}\right)_P = \left(\frac{\partial H_2}{\partial T}\right)_P - \left(\frac{\partial H_1}{\partial T}\right)_P = C_{P_2} - C_{P_1} = \Delta C_P \qquad (3.8)$$

Nothing more is needed beyond the recognition of H as an extensive property. The second equality follows from the definition of C_P (equation (3.2)), the the third equality acknowledges that it also is an extensive property. Any further 'proof' is redundant and, indeed, to be deplored.

Since the law is in the form of a partial differential equation, its practical application requires an integration, calling for knowledge of all C_P terms concerned as functions of temperature. This is why it is handy to have all the empirical equations for the molar heat capacities of pure substances all of the same form, so that appropriate algebraic summation of the coefficients of like powers of T will give the required expression for ΔC_P.

The procedure, for a given reaction, is first to assemble the molar C_P° equations for all the reactants and products. Each equation for a product is multiplied, term by term, by the number of moles of that product produced by the reaction. The resulting equations are added together to give the total heat capacity of the products, ΣC_P°(products). A similar operation is carried out for the reactants, giving ΣC_P°(reactants). ΔC_P° is then calculated:

$$\Delta C_P^{\circ} = \Sigma C_P^{\circ}(\text{products}) - \Sigma C_P^{\circ}(\text{reactants}) \qquad (3.9)$$

If all the molar heat capacities are expressed as in equation (3.4), namely, $C_P^{\circ} = \alpha + \beta T + \gamma T^2 \ldots$, this takes the form

$$\Delta C_P^{\circ} = \Delta \alpha + \Delta \beta T + \Delta \gamma T^2 \ldots \qquad (3.10)$$

so that the Kirchhoff equation can be integrated to give

$$\Delta H^\circ = \Delta H_0^\circ + \Delta\alpha T + \tfrac{1}{2}\Delta\beta T^2 + \tfrac{1}{3}\Delta\gamma T^3 \dots \qquad (3.11a)$$

where ΔH_0°, an *integration constant*, requires comment. The *subscript* zero usually implies 'at 0 °K', and, putting $T = 0$, seems to in equation (3.11a). But these equations, in the present context, are not valid below room temperature, so that this *conventionally used* symbol must not be translated at its face value in this connection.

This type of calculation is important enough to warrant a worked example, as follows.

$$H_2(\text{p.g., 1 atm}) + \tfrac{1}{2}O_2(\text{p.g., 1 atm}) = H_2O(\text{p.g., 1 atm});$$
$$\Delta H_{298\cdot15}^\circ = -577979 \text{ cal}$$

$$C_P^\circ(H_2) = 6\cdot9426 - 0\cdot1999 \times 10^{-3}T + 4\cdot808 \times 10^{-7}T^2$$

$$C_P^\circ(O_2) = 6\cdot0954 + 3\cdot2533 \times 10^{-3}T - 10\cdot171 \times 10^{-7}T^2$$

$$C_P^\circ(H_2O, g) = 7\cdot219 + 2\cdot374 \times 10^{-3}T + 2\cdot67 \times 10^{-7}T^2$$
$$= \text{'}\Sigma\, C_P^\circ \text{ (products)'}$$

These equations are valid from 300 °K to 1500 °K. The task is to find an equation expressing $\Delta H^\circ(T)$ over this range.

There is only one mole of one product, but a summation is required for the reactants, and we need:

$$C_P^\circ(\tfrac{1}{2}O_2) = 3\cdot0477 + 1\cdot6267 \times 10^{-3}T - 5\cdot086 \times 10^{-7}T^2$$

therefore

$$\Sigma C_P^\circ \text{ (reactants)} = 9\cdot9903 + 1\cdot4268 \times 10^{-3}T - 0\cdot278 \times 10^{-7}T^2$$

therefore

$$\Delta C_P^\circ = -2\cdot771 + 0\cdot947 \times 10^{-3}T + 2\cdot95 \times 10^{-7}T^2$$

therefore

$$\Delta H^\circ = \Delta H_0^\circ - 2\cdot771T + 0\cdot4735 \times 10^{-3}T^2 + 0\cdot983 \times 10^{-7}T^3$$

Solving for ΔH_0°

$$-57797\cdot9 = \Delta H_0^\circ - 826\cdot2 + 42\cdot1 + 2\cdot6$$
$$= \Delta H_0^\circ - 781\cdot5$$

therefore

$$\Delta H_0^\circ = -57016\cdot4 \text{ cal}$$

The required equation is therefore

$$\Delta H^\circ = -57016\cdot4 - 2\cdot771T + 0\cdot4735 \times 10^{-3}T^2 + 0\cdot983 \times 10^{-7}T^3$$

If equation (3.5) is used for $C_P^\circ(T)$, the analogue of equation (3.11a) is

$$\Delta H^\circ = \Delta H_0^\circ + \Delta aT + \tfrac{1}{2}\Delta bT^2 - \Delta cT^{-1} \qquad (3.11b)$$

The required equation is therefore

$$\Delta H^\circ = -56992 - 2{\cdot}80T + 0{\cdot}59 \times 10^{-3}T^2 - 0{\cdot}08 \times 10^5 T^{-1}$$

Care is necessary not to overestimate the accuracy of these 'working equations'.

3.6   The formation of gaseous molecules from gaseous atoms

There was no choice in classical thermochemistry other than to adopt as standard states for elements and compounds their respective normal states (solid, liquid or gaseous) at agreed, standard values of temperature* and pressure. This served the immediate purposes, and still does in 'utilitarian thermodynamics'. It does not, however, suit the purpose of fundamental chemical enquiry. Information about the energetics of *molecule formation*, clearly latent in standard enthalpies of formation of compounds, is obscured by extraneous factors, variable in kind and incidence. Polyatomic elements, solid, liquid or gaseous, are used up in the formation reactions, so that energy is consumed in bond-breaking, and in overcoming cohesional forces in condensed phases. When solid or liquid compounds are formed, the satisfaction of intermolecular attractive forces liberates energy, and this is added to the energy release of bond-making. There is, in general, an adventitious mixing together of endo- and exothermic, intra- and intermolecular 'breaking and making' processes that defies immediate interpretation. There is clearly another disability of classical thermochemistry to be removed.

The first step is to get rid of all solids and liquids, and all chemical bonding between like atoms in the elements. All the compounds must be evaporated to the ideal gas state, and all the elements must be not only evaporated but *atomised*. This involves the determination of ΔH° values for processes such as C(graphite) $\rightarrow$ C(gaseous atoms) and $\tfrac{1}{2}H_2$(molecular gas) $\rightarrow$ H(gaseous atoms), the former being very difficult, but the latter easy. Once this has been done for all the elements concerned, the rest is but another application of Hess's law.

The formation of n-hexane can be used in illustration. The classical formation reaction can be set out as in the top line of Fig. 3.3, ΔH_1° representing the normal standard enthalpy of formation, obtained from 'heats

* Older data, 18 °C; modern data, 25 °C.

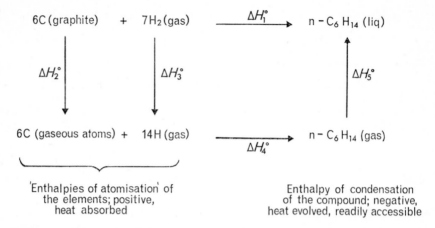

Fig. 3.3 Route to the standard enthalpy of formation of gaseous *n*-hexane from gaseous atoms

of combustion' as already illustrated. The formation can, in principle, be conducted by the alternative route shown; Hess's law then requires that

$$\Delta H_1^\circ = \Delta H_2^\circ + \Delta H_3^\circ + \Delta H_4^\circ + \Delta H_5^\circ$$

This can be solved for ΔH_4°, all the other terms being accessible; it exemplifies the enthalpy of formation of a gaseous compound from its constituent elements in the form of gaseous atoms. A shorter, alternative name sometimes used is 'atomic heat of formation', or, for the reverse, endothermic process, 'heat of atomisation of the gaseous compound'. A generalisation of the whole operation is represented in Fig. 3.4, arranged so that enthalpy increases, non-quantitatively, upwards in the diagram.

The institution of these new standard states (temperature still 298·15 °K; pressure unspecified because ideal gas enthalpy depends only on temperature) clears away the gross irrelevancies. The first result is the disappearance of all endothermic compounds. The questions arise, however, whether the new data are really the functions best suited to theoretical interpretation, and, if so, whether they need yet further refinement? It will pay to consider these questions in some detail because this will give insight into the nature of the chemical problems to be tackled.

Another look at Fig. 3.3 shows that 20 moles of reactant, atomic, gas are consumed to give *one* mole of gaseous, molecular, product. Despite the fact that the entropy contribution of molecules (per mole of molecules) is intrinsically greater than that of atoms (per mole of atoms) because of the greater variety of thermal motions open to them, the formation reaction is

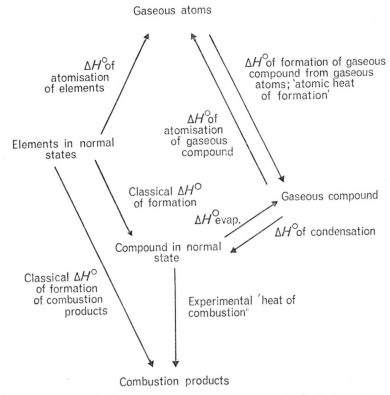

Fig. 3.4 Generalisation of thermochemical route to ΔH° of formation of a gaseous compound from gaseous atoms

attended by a considerably entropy loss—a standing disability of synthetic reactions. This makes ΔG° less negative than ΔH° for the formation reaction $(-\Delta G^\circ < -\Delta H^\circ)$. How does this affect the interests of the theoretical chemists? A little thought suggests that their fundamental interest is really *microscopic*; ultimately they are concerned with the formation of a *bond* by the overlap of bonding orbitals, and do not wish to be confused with considerations of entropy and thermal energy. In this case, they are better off with ΔH° than with ΔG°, and have come back to the Thomsen–Berthelot principle after all.

The question of thermal energy, not inconsiderable at the standard temperature of 298·15 °K remains. It would be better to get rid of it by changing to 0 °K for the standard states—which must, however, remain gaseous (becoming even more hypothetical, but this is no matter). This change can be effected by a Kirchhoff calculation based on heat capacities measured to the lowest accessible temperatures, i.e., by solving

$$\Delta H_0^\circ = \Delta H_{298}^\circ - \int_0^{298} \Delta C_P^\circ \, dT \qquad (3.12)$$

where ΔH_0° now means what it says, and where, by convention, 298 is meant to read 298·15 °K. This step is by no means easy, but for some purposes it is desirable and is taken; the results form part of the body of critically selected data. It should be noted that as $T \to 0$ °K, ΔG° approaches identity with ΔH°, so it might be said that two birds have been killed with one stone.

We still wish to know what difference it makes if this step is *not* taken. This can be assessed in terms of the possible independent contributions to thermal energies. Each of n gaseous atoms, before combining together in the formation reaction, can accommodate energy only in three degrees of freedom of translatory motion; there are thus $3n$ equal contributions to the total thermal energy of the reactants. Modes of motion available to the *one* molecule formed are 3 translational and (for a non-linear molecule) 3 rotational and ($3n - 6$) vibrational. Not all of the latter need be excited (bearing in mind the decreasing 'fine-grainedness' of energy quantisation in the sequence translation, rotation, vibration), nor is it easy to take into account more or less hindered, intramolecular 'free rotations' about single covalent bonds. Although it is therefore impracticable to arrive at a clearcut general summation, ascribing a thermal energy contribution of $\frac{1}{2}RT$ (*ca.* 0·3 kcal mole^{-1}) to each 'squared term', it can be seen that the loss in heat capacity in the formation reaction is offset to an extent increasing with the complexity of the molecule formed. In other words, the effects of ignoring the integral in equation (3.12) will not only be less severe than might at first be guessed, but, over a range of formation reactions, will have some degree of fractional proportionality to ΔH_{298}° itself. The net result is that while ΔH_{298}° always exceeds ΔH_0° (e.g. for $2H = H_2$, 104.18 as against 103·24 kcal), the difference is not large and very seldom causes significant distortions when the main purpose is to make comparisons. ΔH_{298}° for 'atomic formations' is therefore also a critically selected and tabulated function, used in a manner shortly to be illustrated.

Before proceeding, however, it is necessary to ask whether any further 'thermochemical refinement' could be desired, even beyond the stage of ΔH_0°; are there any other energy terms that should be taken into account in relation to the prime interest of chemical bond-making?

There are three such terms. The first arises from the fact, fundamentally based on Heisenberg's uncertainty principle, that molecules at 0 °K retain residual vibrational energy—a *zero-point energy* of half a vibrational quantum for each vibrational mode, namely, $\frac{1}{2}hv_0$ per mode, where h is Planck's constant ($6\cdot626 \times 10^{-27}$ erg sec) and v_0 is the appropriate fundamental vibration frequency. This residual molecular energy ($\frac{1}{2}\sum_i hv_i = 0\cdot0014296\sum v_i$, kcal mole^{-1}, with v in cm^{-1}) is not small; it is about ten times the *removable* thermal energy at 298 °K (for methane, z.p.e. $= 27\cdot1$ kcal mole^{-1}; $H^{\circ}_{298} - H^{\circ}_0$ $= 2\cdot3$ kcal mole^{-1}).[6] This seems shocking, but it must be recognised that zero-point energy is intrinsic to the existence of molecules and cannot be classed as an irrelevance to chemical bonding. It must, however, be noted that it depends not a little on molecular *structure* (n-pentane, 98·5; neo-pentane, 96·8; cyclopentane, 86·0 kcal mole^{-1}) and is not therefore strictly an additive function of the bonds a molecule contains. Correction to ΔH°_0 for zero-point energy has been made to obtain '*binding energy terms*', but the field in which this can usefully be done is limited.[7]

The second energy term included in ΔH°_0 is to do with intramolecular, non-bonding interactions, with origins somewhat hard to identify. Reference must be made to the forces which operate between atoms and molecules which are *not* chemically bonded to each other. They include the London, dispersion forces of attraction, and the very strong forces of repulsion that come into play at closer distances of approach. It is the balance between these which define the *van der Waals radii* of atoms. There is no reason to suppose that similar forces do not act between 'non-bonded atoms' in a molecule—that is, between atoms, not directly bonded to each other, belonging to the same molecule. Those in second, or even third, nearest neighbour juxtaposition are near enough. The interaction of the charge clouds of neighbouring bonds attached to the same atom has also been discussed. However this may be, phenomena such as the rotational barrier in ethane ('free' rotation of the methyl groups has to surmount energy humps of 3 kcal mole^{-1}) leave no doubt as to the significance of such effects.

The third energy term fundamentally affects the significance of ΔH°_0. The gaseous atoms involved in the formation reaction at 0 °K have very reasonably been assigned to their unexcited *ground states* of minimum energy. These, however, are not the *valency states*. Should not 'true' or '*intrinsic bond energies*' be measured relative to valency states rather than to ground states?

It would be beyond our scope to discuss the possible implementation of such a scheme—it is not, for example, just a matter of the distinction between the s^2p^2 and sp^3 states of the carbon atom. The short answer, then, is, that to set up this changed scale would really represent a change of objective, and is in any case hardly practicable except as a theoretical exercise aimed at independent assessment of bond dissociation energies (see later). The valency states of atoms have no independent existence, and the energies required to reach them from the ground states are not directly observable. It can indeed be argued that atoms in valency states are already parts of molecules, and that the processes of promotion, hybridisation and spin redistribution which get them there must logically be considered as part and parcel of bond formation.

This brings attention back from the atoms to the molecules. Their properties depend intimately on the valency states—the kind of hybridisation—of the atoms they contain, in a manner which forbids confinement of attention rigidly to *bonds*. Even the simplest molecule must be viewed as a whole. This can be emphasised by an example.

The water molecule contains two equivalent O—H bonds, but this statement alone is inadequate to explain its properties. The oxygen orbitals are hybridised (approximately sp^3), and two lone pairs of electrons occupy orbital lobes of considerable radial extension away from the hydrogen atoms. In many respects, these lone pairs dominate the behaviour of the water molecule. Lone-pair–lone-pair repulsion is significant in determining the valency angle of $104\cdot5°$, and the 'polarity' of the molecule is derived less from the bonds than from the separation in space of the charge on the nucleus of the oxygen atom and those of its lone-pair electrons. So the very hybridisation which is part and parcel of the bonding has energetic and other implications which it would be hard to incorporate logically in a discussion confined to bonds, as bonds.

To pursue the argument a little further, it becomes obvious that if the promotion of atoms to valency states is apportioned to bond formation, variable valency must be critically significant to bond formation energy. If, for example, we consider PCl_3 (pyramidal) and PCl_5 (trigonal bipyramidal), it would not be reasonable to expect a uniform P—Cl bond energy, either for the two compounds, or indeed for the equatorial and axial bonds in PCl_5.

Reverting to the original purpose of discussion, it can be concluded that any attempt to 'refine' ΔH_0° leads directly into deep waters. Indeed, within the context of uncertainties that have appeared, ΔH_{298}° is, for most purposes, likely to be just as serviceable. But some rather more fundamental considerations have been underlined. *Delocalisation*, in various aspects, cannot in principle be ignored, although this does not prohibit approximations subsequently shown to be reasonably satisfactory for restricted purposes. Fundamentally, however, the formation of a molecule should be regarded as an integrated process involving all the constituent atoms—indeed, the nuclei and electrons. It follows that there is restricted validity in seeking to express the total 'atomic formation energy' of a molecule as a summation of reproducible and precisely quotable bond formation energies. This is important in order to place the two following sections in proper perspective.

3.7   Thermochemical bond energy terms

It is convenient to consider heats of atomisation of gaseous molecules (ΔH_a° for short), rather than the long-winded 'enthalpies of formation of gaseous molecules, etc., etc.'; in other words, let $\Delta H_a^\circ = -\Delta H_{298}^\circ$ for the time being. Then, for atomisation of a diatomic molecule, ΔH_a° is identical with the bond dissociation energy, frequently symbolised by D, and expressed in kcal mole^{-1}, i.e.

$$XY(g) = X(g) + Y(g); \quad D_{X-Y} = \Delta H_a^\circ \tag{3.13}$$

For a molecule XY_n, having n bonds of the same kind, the total heat of atomisation can be apportioned equally:

$$XY_n(g) = X(g) + nY(g); \quad E_{X-Y} = \Delta H_a^\circ / n \tag{3.14}$$

where E_{X-Y} is a *bond energyterm*. This is not identical successive bond dissociation energies ($D = \Delta H_1, \Delta H_2 \ldots \Delta H_n$) for the *stepwise* removal of Y atoms, although, by Hess's law, they must add up to the total ΔH_a° for complete atomisation. On the contrary, the real situation is as shown in Table 3.1.

Table 3.1   Bond dissociation energies, ΔH_1, ΔH_2 ... , and thermochemical bond energy terms for XY_n, kcal mole^{-1}

XY_n	ΔH_1	ΔH_2	ΔH_3	ΔH_4	E_{X-Y}
CO_2	127	256	—	—	192
OH_2	118	102	—	—	110
$HgCl_2$	81	25	—	—	53
CH_4	102	88	124	80	99
$TiCl_4$	80	101	106	124	103
$AlCl_3$	91	95	119	—	102

In the first example, breakage of one of the two equivalent C=O bonds in carbon dioxide transforms the other into the rather stable triple bond of carbon monoxide (cf. N_2; $\Delta H_a^\circ = 226$ kcal mole^{-1}). The *general* explanation is that each one-at-a-time removal of an atom Y alters the valency state of the atom X from which it is detached, and the relative values of the bond dissociation energies are susceptible to theoretical explanation. For instance, the loss of one chlorine atom from $HgCl_2$ has but a minor effect on the Hg atom; removal of the second, on the other hand, causes reversion from some kind of sp hybridisation to the s^2 ground state. The second step releases substantial energy which offsets that required to break the bond.[8]

It is therefore clearly important not to confuse thermochemical bond energy terms, always functions of complete atomisation, and bond dissociation energies. The latter are functions of specific bonds, broken one at a time in specified molecules or radicals; they are determinable directly by spectroscopic and electron impact methods.

On turning attention to polyatomic molecules containing more than one kind of bond, a new difficulty appears: there is only one heat of atomisation, ΔH_a° per molecule, and no unique way of dividing it into different bond energy terms. The only recourse is to the assumption made by Fajans in 1920—that for a given kind of bond, the bond energy is constant from one molecule to another. Then ΔH_a° data for two or more compounds will give simultaneous equations soluble for the required unknowns. For instance, the heat of atomisation of an alkane, C_nH_{2n+2} can be expressed as a sum of bond energy terms:

$$\Delta H_a^\circ = (n-1)E_{C-C} + (2n+2)E_{C-H} \qquad (3.15)$$

values of ΔH_a° for any two homologues will give the desired results. Table 3.2 shows that this pans out very well for higher, but not for lower, normal

Table 3.2 Bond energy terms from $-\Delta H_{298}^\circ$ for n-alkanes, kcal mole^{-1} (based on heats of atomisation : C, 170·9; H, 52·09 kcal mole^{-1})

Alkane	E_{C-C}	E_{C-H}
CH_4	—	99·29
CH_4/C_2H_6	78·84	99·29
:	:	:
:	:	:
C_5H_{12}/C_6H_{14}	82·82	98·60
C_6H_{14}/C_7H_{16}	82·72	98·64
C_7H_{16}/C_8H_{18}	82·70	98·64
C_8H_{18}/C_9H_{20}	82·63	98·68

alkanes. Larger discrepancies appear if branched chain alkanes are brought in—only partly to be accounted for by differences in zero-point energy. In this respect, therefore, the principle of the additivity of bond energy terms fails, which should not occasion surprise.

A more sensitive test is found in 'redistribution reactions'; they are reactions involving no change in kind or number of bonds, and should be athermal, i.e. have $\Delta H^\circ = 0$. Two examples are

$$\tfrac{3}{4}CCl_4 + \tfrac{1}{4}CF_4 = CCl_3F; \quad \Delta H^\circ = 8 \cdot 9 \text{ kcal}$$

$$\tfrac{3}{4}C(CH_3)_4 + \tfrac{1}{4}CCl_4 = (CH_3)_3CCl; \quad \Delta H^\circ = -6 \cdot 9 \text{ kcal}$$

So the additivity principle fails again, quite badly. This raises a question of principle, met also in other physico-chemical fields, which warrants brief comment.

Deviations are found from a general rule which has been established as true to a certain level of approximation. There are two alternative policies. The first is to systematise the deviations and seek valid explanations for them. This can be done in the alkane field by adding 'constitutional' terms to the right-hand side of equation (3.15) to allow, for example, for specific, pair-wise interactions of non-bonded atoms, or other factors of possible significance. Provided this does not add too many undetermined parameters, and sufficiently numerous data are brought into a proper statistical analysis, this is justifiable and fruitful. The hydrocarbons lend themselves to this procedure,[9] which, in general, has quite wide possibilities in organic chemistry (where a lot of chemistry is based on *few* elements). There are many variations of this kind of systematisation, bringing in the other extensive thermodynamic properties,[10] but this is beyond the immediate scope of discussion.

The other policy is to accept the rule for what it is worth, and use it within its limitations; broadly, this is all that can be done in inorganic chemistry (where a lot of chemistry is based on *many* elements) because the conditions permitting finer analysis seldom appertain.

Heats of atomisation of the elements are basic to either application; some examples from a recent critical publication[11] are collected in Table 3.3.

Table 3.3 Heats of atomisation of elements, ΔH°_{298}, kcal (g atom)$^{-1}$

H	52·10	P	75·59
C	170·91	S	65·23
N	112·98	Cl	28·95
O	59·56	Br	26·73
F	19·0	I	25·48

The only unambiguous bond energy terms are those relating to diatomic molecules—some are shown in Table 3.4. It is of interest, as can be seen from the tables, that fluorine forms a more stable bond with another halogen than with itself. The comparative instability of F_2 is attributed to repulsions between non-bonding electrons, acting across the short bond.

Table 3.4   Bond energy terms, $-\Delta H^{\circ}_{298}$, kcal mole^{-1} for diatomic molecules

HF	135·3						
HCl	103·12	FCl	59·8				
HBr	87·6	FBr	55·8	ClBr	52·2		
HI	71·3	FI	67	ClI	50·5	BrI	42·5

It is not a rewarding exercise to try to assemble a table of bond energy terms covering a wide range of elements. The variations from one source to another are large, and this is undoubtedly mainly due to failure of the additivity rule. The usual policy is therefore to rely on the intuition of inorganic chemists; Table 3.5 has therefore been taken from an authoritative text.[12] The appended data for multiple bonds are taken from Pauling.[13]

Table 3.5   Representative single bond energy terms, $-\Delta H^{\circ}_{298}$, kcal mole^{-1}

	H	C	N	O	F	Si	P	S	Cl	Br	I	
H	104	99	93	111	135	70	76	81	103	88	71	H
C		83	70	82	116	69	63	62	79	66	57	C
N			38	48	65	—	(50)	—	48	58?	—	N
O				33	44	88	(84)	—	49	—	48	O
F					37	129	117	68	61	57	—	F
Si						42	(51)	54	86	69	51	Si
P							41	(55)	76	65	51	P
S								63	66	51	—	S
Cl									58	52	50	Cl
Br										46	43	Br
I											36	I

Multiple bond energy terms, $-\Delta H^{\circ}_{298}$, kcal mole^{-1}:

C = C	147	C = S	114
N = N	100		
C = N	147	C ≡ C	194
C = O	164 in formadehyde	C ≡ N	207 in hydrogen cyanide
	171 in other aldehydes		213 in nitriles
	174 in ketones		

3.8   Some applications

If the additivity rule holds, heats of atomisation of molecules of known structure, calculated by adding up appropriate bond energy terms, should agree with those determined experimentally, i.e. derived from heats of combustion by the kind of Hess's law calculation illustrated in Fig. 3.3. In other words, $-\Delta H^{\circ}_{298} = \Sigma E$ should be satisfied. For very numerous organic compounds this is found to be so, give or take a kilocalorie or two. Cases of large discrepancies therefore warrant special attention, and are usually due to a mistake in the 'known structure'. If, for example, the classical Kekulé structure is ascribed to benzene, so that the aromatic ring is represented as containing alternating $C - C$ and $C = C$ bonds, E is $6E_{C-H} + 3E_{C-C} + 3E_{C=C} = 1283$ kcal mole^{-1}. On the other hand, $-\Delta H^{\circ}_{298}$ for benzene (from its heat of combustion) is 1324 kcal mole^{-1}. The discrepancy of 39 kcal mole^{-1}, although subject to considerable uncertainty, is large, and is in the sense to show that the benzene molecule is more stable (more heat is liberated when it is formed from its constituent atoms) than could be expected from its formulation as cyclohexatriene. This is the extra aromatic stability attributed in valence bond theory to *resonance*, and the discrepancy becomes the 'resonance energy'. The Kekulé formulations are fictitious, but convenient, *canonical forms* regarded as contributing to the real state of the benzene molecule.

This is, however, a poor, if versatile, method of estimating such an energy; it comes out as a difference between two much larger quantities—one being of very questionable accuracy. Whenever possible it is better to get rid of ΣE and, instead, to make comparison between the ΔH°_{298} data for suitably chosen reactions other than combustion. Kistiakowsky has adapted hydrogenation reactions to accurate calorimetry, and has compared, for example, the heats of hydrogenation of benzene and of cyclohexene (six-membered ring, 5 C—C bonds and *one* undoubted C=C bond). The results were:

$$C_6H_{10} + H_2 = C_6H_{12}; \ \Delta H^{\circ}_{298} = -28 \cdot 59 \text{ kcal}$$

$$C_6H_6 + 3H_2 = C_6H_{12}; \ \Delta H^{\circ}_{298} = -49 \cdot 80 \text{ kcal}$$

If benzene were cyclohexatriene, the expected heat of hydrogenation would be $\Delta H^{\circ}_{298} = 3(-28 \cdot 59) = -85 \cdot 77$ kcal. Again it appears that the benzene molecule lies at a lower energy level than would a molecule of Kekulé structure, this time by $35 \cdot 97$ kcal mole^{-1}, which is clearly a better estimate of the resonance energy. Pauling has listed numerous empirical resonance energy values obtained thermochemically.[13]

Although it is impracticable to explore further into this mainly organic field, it is clear in principle that differences between observed ΔH°_{298} data and those 'expected' (usually on better grounds than ΣE) can be informative on any phenomenon that affects molecular stability—resonance, hyperconjugation (delocalisation), steric hindrance, steric strain.*

* The 'strain energy' of tricyclo-octane, 'cubane', C_8H_8 has recently been estimated as 157 kcal mole^{-1}.

3.9   Electronegativities

Valence bond theory allows a covalent bond to be imagined in terms of resonance between covalent and ionic forms. For a symmetrical diatomic molecule, $A-A$, the contributions of oppositely polarised ionic forms, $A^+ A^-$, $A^- A^+$ are small and equal. Pauling[13] calls such a hybridised bond of minimised total energy content a *normal covalent bond*. The same will be so for another symmetrical molecule, $B-B$, and *may* also be the case for the unsymmetrical molecule $A-B$. If this situation exists, Pauling suggests that the bond energies should be related by*

$$D_{A-B} = (D_{A-A} \cdot D_{B-B})^{\frac{1}{2}} \tag{3.16}$$

If, on the other hand, one ionic form, say $A^+ B^-$, makes a greater contribution than the other, $A^- B^+$, the bond can be said to have ionic character, and is additionally stabilised by loss of 'ionic resonance energy'. In this case, the equality of equation (3.16) will not apply; instead,

$$\Delta = D_{A-B} - (D_{A-A} \cdot D_{B-B})^{\frac{1}{2}} \tag{3.17}$$

where Δ is always of a sign to represent additional stability of the bond A—B, i.e. always positive if bond energy is counted positive.

It would be convenient if Δ could be expressed as a function of characteristic properties of the atoms A and B *in their covalently bound states*. This led to the definition of *electronegativity*† as 'the power of an atom in a molecule to attract electrons to itself'. Then Δ for the bond A—B should be some function of the difference between the electronegativities of the atoms A and B. The desired function was found, by exploration, to be

$$x_A - x_B = k\Delta^{\frac{1}{2}} \tag{3.18}$$

where $x_A - x_B$ is the electronegativity difference between the A and B (expressed in electron volts, eV; 1 eV atom^{-1} = 23·06 kcal mole^{-1}), Δ is in kcal mole^{-1} and k is an appropriate constant. Absolute values of x were arbitrarily adjusted to provide a scale of figures ranging from 2·5 for carbon

---

* The geometric mean works out better than his earlier try-out of the arithmetic mean.
  † As will be seen, not to be confused with electron affinity, and certainly not with any normal electrochemical concept.

to 4·0 for fluorine. If this is valid, it provides a means for estimating 'awkward' bond energy terms by use of

$$D_{A-B} = (D_{A-A} \cdot D_{B-B})^{\frac{1}{2}} + 23 \cdot 06(x_A - x_B)^2 \qquad (3.19)$$

The bracketed figures in Table 3.5 were obtained in this way.

In view of the somewhat shaky status of bond energy terms at large, the justification, or even point, of this exercise might seem questionable. The results are, however, supported by alternative methods of estimating electronegativities.

Mulliken (1935) pointed out that the energy to form ion pairs A^+, B^- or A^-, B^+ from atoms A, B is equal to the difference between appropriate atomic ionisation potentials, I, and electron affinities, E. If these properties are not significantly altered in the molecule A—B, a normal covalent bond would result from the equality $I_A - E_B = I_B - E_A$. If, on the other hand, A were more 'electronegative' than B, then $(I_A - E_B) > (I_B - E_A)$, leading to partial ionic character in the sense $A^- - B^+$. Rearrangement of the inequality gives $(I_A + E_A) > (I_B + E_B)$. Hence, the average of ionisation potential and electron affinity should give a measure of electronegativity. Values so obtained stood in quite good linear relation with Pauling's figures.

Another method due to Allred and Rochow (1958),[14] considered the electronegativity of a bonded atom to be a function of the electrostatic attraction exerted by the nuclear charge on the bonding electrons, situated at a distance from the nucleus equal to the covalent radius of the atom. These electrons 'see' only an 'effective' nuclear charge because of the screening effect of the inner electrons; this effect is calculable by well-established rules. The results also were linearly related to Pauling's.

In either case, the best straight lines were used to get the data on to the Pauling scale. The agreement between thermochemically derived electronegativities and those from quite different sources (more recently supplemented by an ingenious application of nuclear magnetic resonance)[15] must be considered impressive, lending credence to the significance of the concept. This must not, however, afford protection from criticism such as the following.

Agreement is best between Pauling and Allred–Rochow figures; $\pm 6\%$ for the first 30 elements. They have, of course, been arbitrarily adjusted to agree as well as may be, and all contain an added constant term; in any case it is normally only differences between pairs of them that are any use. Very much larger discrepancies occur later in the periodic table. An electronegativity of a given element can relate only to *one* valency state. Data for alternative states of transition elements, and others of variable valency, are not available, or very likely to become available. The concept of 'percentage ionic character', ostensibly calculable from electronegativity difference, is to be regarded as a first approximation, now passing out of modern valency theory. It is dangerous to use electronegativity difference to deduce anything about the *polarity* of a bond; the C—I 'bond moment' is 1·64 D*, although carbon and iodine have nearly identical electronegativities. Despite all this, it still appears that electronegativity is a usefully informative *parameter* which remains part of the lore of modern inorganic chemistry. A selection of electronegativities (conservative in 'accuracy') is appended in Table 3.6.

Table 3.6 Selected electronegativities of some elements in normal valency states

H	2·1										
Li	1·0	Be	1·5	B	2·0	C	2·5	N	3·0	P	3.5	F	4·0
Na	1·0	Mg	1·2	Al	1·5	Si	1·8	P	2·1	S	2·5	Cl	3·0
K	0·9	Ca	1·0	Ga	1·8	Ge	2·0	As	2·2	Se	2·5	Br	2·8
Rb	0·9	Sr	1·0							Te	2·0	I	2·4

3.10 Ionic compounds

Localised bonds, and bond energies, are not appropriate to electrovalent substances—unless for gaseous ion-pairs. The forces responsible for the cohesion of ionic crystals are electrovalent and no longer irrelevant to fundamental chemical interests. There is therefore, particularly in the inorganic field, a wider application of Hess's law, bringing in energy terms yet to be considered. There are the so-called '*ionisation potentials*' and '*electron affinities*', and the lattice energies of crystals. Each needs brief discussion.

First, second, ... nth ionisation potentials, $I_1, I_2 ... I_n$, eV atom^{-1}, are the names and units that by traditional and common usage express the energies *required* to effect successive processes $X(g) = X^+(g) + e^-(g)$, $X^+(g) = X^{2+}(g) + e^-(g)$ and so on, at 0 °K. They are mostly accurately determinable from the series limits of atomic spectral lines, less accurately by other methods (e.g. electron impact, surface ionisation on hot tungsten). For a given atom, ionisation potentials increase in sequence $I_1, I_2 ... I_n$

* D = Debye unit = 10^{-18} e.s.u. Asymmetry of charge arising from electronegativity difference is *one* factor determining bond moment. Others are, the 'homopolar' dipole (arising from unequal sizes of the orbitals of the two atoms forming the bond), the asymmetry of bonding orbitals (and of lone pair orbitals) arising from hybridisation. This re-emphasises that a molecule must be viewed as a whole.

because of the obviously increasing coulombic work of separating opposite electrical charges to infinity. They are functions of the effective nuclear charge 'seen' by the ionising electron, and of the radius of the atom or ion it is leaving; both of these are dependent on the screening of the nucleus by the filled, or partially filled, s, p, d and f shells. Trends of ionisation potentials in the periodic table accordingly reflect, and are evidential of, features of atomic structure. This is a matter so basic in interest to chemistry that illustration of it cannot be omitted.

First ionisation potentials are periodic functions of atomic number (as fragmentarily shown in Fig. 3.5a) with maxima at the noble gases, with their very stable $1s^2$ or ns^2np^6 outer shells, and minima at the alkali metals, consistent with the ease of removal of the unpaired ns electron. In lithium, this outer electron is well shielded by the $1s^2$ shell; in succeeding alkali metals, radii increase, and although shielding by ns^2np^6 shells is somewhat less effective, first ionisation potentials fall (Fig. 3.5c). The relative ease of removal of an unpaired electron is shown by the fall of first ionisation potential from beryllium to boron. The similar fall from nitrogen to oxygen is due to a degree of stability associated with the half-filled p orbitals $(2p_x^1 2p_y^1 2p_z^1)$ in O^+. It is clear from Fig. 3.5a that $I_1, I_2 \ldots I_n$, plotted as functions of atomic number, show similarities—but displaced along the scale of abscissae by one unit for each electron removed. This is to be expected if each ion of positive charge n is isoelectronic with the atom or ion of immediately preceding atomic number and charge $(n - 1)$. It is also to be expected that the jump from I_n to I_{n+1}, where n is group number, is large.

The transition series of elements present features of interest (Fig. 3.5b). From Sc to Zn, each increment of positive nuclear charge is rather poorly screened by the simultaneously added $3d$ electrons, but it is understandable that while the first ionisation potentials increase along the series, they are on the low side and much of a muchness. A more important feature requires more detailed analysis.

The energy of the $3d^n$ shell of electrons is a function of their coulombic attraction to the screened nuclear charge (i.e. to the Ar atomic core), and of interaction between the d electrons. There are two contributions to this interaction. The first is coulombic repulsion—destabilising; the second is a non-classical exchange energy, proportional to the number of pairs of parallel spins, m—it is stabilising in effect. Considering the filling of the $3d$ orbitals *in the dipositive ions* from $3d^1$ to $3d,^{10}$ remembering the maximisation of unpaired spins (Hund's first rule) it is easy to tot up the m values. They are, successively 0, 1, 3, 6, 10, 10, 11, 13, 16, 20; this explains why the I_3 values run as shown in Fig. 3.5b.

The poor screening provided by the $(n - 1)d^{10}$ sub-shell for a lone ns

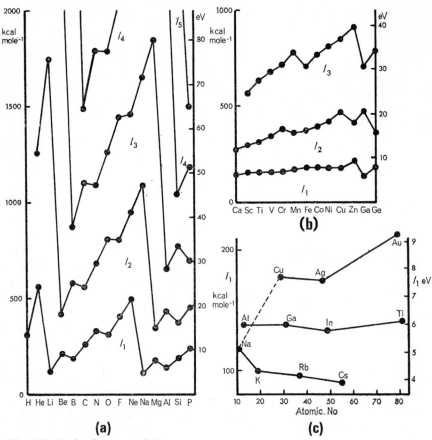

Fig. 3.5   Ionisation potentials
   a. successive ionisation potentials of non-transition elements
   b. successive ionisation potentials of some transition elements
   c. some ionisation potentials reflecting sub-shell screening

electron (as compared with that of $(n - 1)s^2(n - 1)p^6$) is shown (Fig. 3.5c) by the first ionisation potentials of Cu([Ar]$3d^{10}4s$), Ag([Kr]$4d^{10}5s$) and Au([Xe]$5d^{10}4s$). Gold also shows the effect of the lanthanide contraction. This defect in screening is, however, rectified in $(n - 1)d^{10}ns^2$, as can be seen in the fall of I_1 from Zn to Ga; In and Tl also have about the same low I_1 for a similar reason, and consequently tend to show resemblances, in their univalent states, to Rb, Cs and Fr (Fig. 3.5c). This is, of course, the 'inert pair' effect responsible for the increasing relative stabilities of the bivalent states of Ge, Sn and Pb.

It will become yet more clear that ionisation potentials (of molecules as well as atoms) are a great gift in more than one field of enquiry, and one can hardly refrain from mentioning the following dramatic application.

Bartlett,[16] studying an oxyfluoride of platinum, PtO_2F_6, showed that it was isostructural with $KSbF_6$ and must be dioxygenyl hexafluoroplatinate (V), $O_2^+[PtF_6]^-$. He then synthesised it from oxygen gas and PtF_6 at room temperature. He observed that I_1 for O_2 (12·2 eV) is almost identical with that of Xe (12·13 eV), so tried the obvious experiment. This gave an orange solid, subliming on heating in vacuum; $XePtF_6$,* the first true compound of an 'inert gas'. This removed a traditional inhibition, and initiated a new chapter of chemistry in which there has been intense activity over the last few years.

Electron affinities relate to processes $X(g) + e^-(g) = X^-(g)$; $X^-(g) + e^-(g) = X^{2-}(g)$, and are traditionally expressed as energy *released*, in eV atom^{-1}. Surprisingly, they present a very different picture. They are difficult to measure; electron impact, photo-ionisation, surface ionisation on hot tungsten are the independent, direct methods, but they are limited in application and accuracy. Supplemented by quantum-mechanical calculation, extrapolation from isoelectronic series (Fig. 3.5a shows the basis; I_1 of X is of course the same as the electron affinity of the ion X^+), they provide but a sparse and somewhat discrepant list,[11] so that recourse is usually had to deriving electron affinity from a Hess's law calculation of a kind shortly to be illustrated. Table 3.7 contains a few data. The values for halogen atoms are not as might be expected from their electronegativities (Table 3.6); the

Table 3.7   Some electron affinities, expressed as $\Delta H°$ for $X(g) + ne^- = X^{n-}$, kcal mole^{-1}

H	−17	S	ca. − 48
O	−34	S(2e$^-$)	ca. + 100
O(2e$^-$)	ca. + 170	Cl	− 87
F	−82·5	Br	− 82
Si	−34	I	− 76
P	−18		

reluctance of oxygen to take on a second electron is also noteworthy and perhaps surprising.

Electron affinities of molecules and radicals are also tabulated.[11] It is worthy of comment that the stability of $XePtF_6$ is attributable to the

* Later X-ray study showed the constitution to be somewhat more complex, but this does not affect the argument, nor the glory.

undoubtedly large, if undetermined, electron affinity of PtF_6. This seems to be a common feature of similar real, or hypothetical molecules, revealed in the stabilising effect of the corresponding anions (with electron affinity satisfied) on unusual cations,[17] such as in the compounds $O_2^+BF_4^-$, $Cl_2F^+BF_4^-$, $ClO_2^+BF_4^-$, $Xe^+SbF_6^-$, $NF_4^+SbF_6^-$.

The *lattice energy* of an ionic, crystalline substance is the energy required to transform one mole of it at $0\ °K$ into non-interacting gaseous ions. It is usually symbolised by U_0; then $U_0 = \Delta H_0^\circ$ for the generalised process

$$M_m^{n+}X_n^{m-}(c) = mM^{n+}(g) + nX^{m-}(g) \tag{3.20}$$

Alternatively, the lattice energy may be regarded as the energy released when the infinitely separated ions, devoid of thermal motion, come together to occupy their proper positions in the stable crystal lattice. At $0\ °K$, these positions will be such as to minimise total energy, with the appropriate balance struck between the interionic forces of attraction and repulsion. If, however, the process takes place at a finite temperature, the energy–entropy balance must be taken into account, and the structure of the crystal will minimise the Gibbs free energy. The large entropy loss accompanying the condensation of the gaseous ions increases in significance with rising temperature, with consequences already discussed. This is not our concern at present—whatever the structure, the interest is the lattice *energy*, and it is the safest policy to isolate this purely energetic term by conducting our hypothetical operations at $0\ °K$.

As in a previous case, the question of course arises, how far it may be justifiable to use $298\cdot15\ °K$ as the standard temperature, instead of the inconvenient $0\ °K$?

To examine this, consider the transformation represented by equation (3.20) conducted at constant temperature and pressure; T, P. As before, the Kirchhoff theorem indicates that

$$\Delta H_T^\circ = \Delta H_0^\circ + \int_0^T \Delta C_P^\circ \, dT \tag{3.12}$$

which, for this case, may be expanded in the following way.

The products of the transformation—the separated ions—behave by definition in the same way as $(m + n)$ moles of ideal gas. For one mole of ideal gas,

$$C_P^\circ = \tfrac{3}{2}R + R \tag{3.21}$$

where R is the gas constant. It is the sum of two terms. The first is the rate of increase with temperature of the kinetic energy in three independent modes of translatory motion; the second represents the additional energy required for the work of expansion at the constant pressure, P. Applying this to the transformation in question,

$$\Delta H_T^\circ = \Delta H_0^\circ + \int_0^T (m + n)\tfrac{3}{2}R \, dT + \int_0^T (m + n)R \, dT - \int_0^T C_P^\circ(c) \, dT$$

where $C_P^\circ(c)$ stands for the molar heat capacity at constant pressure of the crystalline solid $M_m^{n+}X_n^{m-}$. Then, since R is a fundamental constant,

$$\Delta H_T^\circ - (m + n)RT - \Delta H_0^\circ = (m + n)\tfrac{3}{2}RT - \int_0^T C_P^\circ(c) \, dT$$

i.e. $$U_T - U_0 = (m + n)\tfrac{3}{2}RT - \int_0^T C_P^\circ(c) \, dT \qquad (3.22)$$

From this it is seen that U_T, the lattice energy at the finite temperature T is identified with the enthalpy of the transformation of the lattice to gaseous ions, ΔH_T°, *less* the 'PV contribution' to the enthalpy of the gaseous ions. The like component of the enthalpy of the solid is negligibly small, and is retained *in* the final integral of equation (3.22)—which is itself, however, by no means negligible. Bearing in mind that RT is about 0·6 kcal mole^{-1} at 25 °C, it can be seen that the significance of the 'PV term' is not great; indeed, because of compensation between thermal energies (vibrational in solid,* translational in gas), U_{298} seldom exceeds U_0 by more than a few kcal— often no more than the uncertainties of other quantities combined with lattice energies in thermochemical calculations. Many, but not all, such calculations can therefore profitably be done in terms of U_{298}.

How lattice energy, U, with ionisation potential, I, and electron affinity, E_a, fits into the thermochemical scheme of things is indicated in Fig. 3.6, which it is useful to compare with Fig. 3.4. Interesting as each thermochemical quantity may be in its own right, the greatest benefit is obtained when all the quantities required to complete what is usually called a *cycle* become available. In terms of Fig. 3.6, for example, we can proceed from the elements in their normal states directly to the ionic, crystalline compound. Alternatively, we can start with the same initial system, atomise, ionise, form the lattice and finish with the same final system, thus completing a Hess's law 'loop'. The ΔH_0° must be identically the same for either path. Considering a cyclic sequence of changes, starting and finishing at the same system, the

* Recollect the classical law of Dulong and Petit.

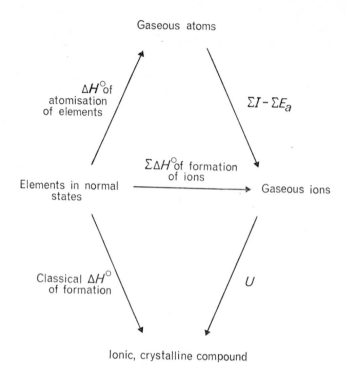

Fig. 3.6 A thermochemical cycle

sum of the ΔH_0° values for the changes must add up to zero. If any *one* of them is unknown, it can be calculated from the others. But the interest and importance of such cycles is much greater than this for the following general reason.

A given crystalline, ionic compound is stable with respect to, and is formed exothermally from, its constituent elements in their normal states. This is a flat statement of fact, associated with one step in a cycle—the lower, left-hand step of Fig. 3.6. If, however, we have knowledge of the other three steps in the cycle, constituting the other route by which the compound can be formed from its elements, we can begin to get at the reasons *why* the compound is stable in terms of the fundamentals of atomic structure and solid state structures. Conversely, we can profitably seek reasons why other compounds are not stable with respect to their elements or to other decomposition products. In short, the completion of thermochemical cycles opens the way to a wide field of information essential to the better understanding of, especially, inorganic chemical facts. The reader might remark that, as to questions of stability, he thought that $-\Delta G^\circ$ of formation was the only valid basis of assessment. This is true, but at $0\,°K$, $-\Delta H_0^° = -\Delta G_0^°$. At $298\cdot15\,°K$ (but not at high temperatures), $\Delta H_T^°$ normally remains safely the major contribution to $\Delta G_T^°$, and, in any case, the first contribution to examine. It is a matter of proportion, and, where necessary, Gibbs free energy cycles can, in principle, equally well be traced.

The classical *Born–Haber* cycle (1919) can be illustrated, in terms of an ionic compound MX_n, as in Fig. 3.7, where the symbols are those already

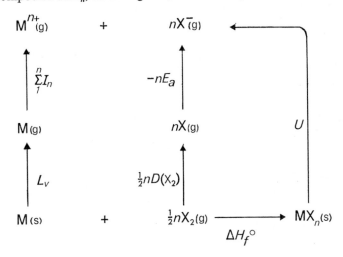

Fig. 3.7 The Born–Haber cycle

used, or are obvious in meaning. For simplicity, temperature is unspecified, and PV contributions ignored. The cycle can be used to calculate lattice energy, provided all the other terms are known, by

$$U = L_V + \sum_1^n I_n + \tfrac{1}{2}nD(X_2) - nE_a - \Delta H_f^°(MX_n)\qquad(3.23)$$

It is of interest to put figures to this for specific examples. For MgO, they are, in the same order (in kcal)

$$U = 36 + 523 + 59 + 170 + 144 = 932$$

If one looked at the large energies *required* to generate Mg^{2+} ions, to dissociate O_2 molecules and to doubly charge oxygen atoms, one might suppose MgO to be a rather unstable oxide. It is, of course, very stable, melting above $2500\,°C$. Its healthy standard enthalpy of formation ($\Delta H_{298}^° = -143\cdot77$ kcal mole^{-1}; cf. $\Delta G_{298}^° = -136\cdot12$ kcal mole^{-1}) is seen to be due to its huge lattice energy, reflecting the very powerful forces of attraction between doubly, oppositely charged ions, assembled in a rock-salt lattice, with a nearest neighbour (internuclear) distance of 2·10 Å.

Lithium fluoride, with almost the same standard enthalpy of formation, presents quite a different picture; the data, arranged in the same sequence are

$$U = 38 + 124 + 19 - 83 + 146 = 244 \text{ kcal}$$

Here the main contributions to the stability of the compound with respect to its elements are the low ionisation potential of lithium, the weak bond in the fluorine molecule, and the high electron affinity of the fluorine atom. The crystal structure is the same (rock-salt) as that of magnesium oxide, and the nearest neighbour distance (2·01 Å) is slightly less; the ratio of the lattice energies (MgO/LiF = 3·82) is seen to be consistent with the operation of Coulomb's law. This gives a clue as to the independent calculation of lattice energies, greatly to be desired, so that they can be put *into* the Born–Haber cycle, as well as got out—in the latter case with the restriction of shaky or inaccessible electron affinities. The calculations are made on the following lines.

The potential energy of a pair of ions of opposite charges Z_+e and Z_-e (valency times magnitude of electronic charge) at a distance r apart is $-Z_+Z_-e^2/r$, relative to zero energy at infinite separation. To obtain the total electrostatic energy of a space-lattice of ions, a summation must be made of many such terms because each ion exerts a Coulombic force on all the others. In the familiar rock-salt structure, each ion interacts with 6 equivalent nearest neighbours of opposite charge at a distance r_0, with 12 next-nearest neighbours of the same charge at a distance $2^{\frac{1}{2}}r_0$, and so on throughout the crystal; attraction and repulsion alternate, making opposite contributions to the total energy. For the rock-salt lattice, the energy of interaction between one ion and all the rest comes to

$$\frac{-Z_+Z_-e^2}{r_0}\left(6 - \frac{12}{\sqrt{2}} + \frac{8}{\sqrt{3}} - \frac{6}{\sqrt{4}} + \frac{24}{\sqrt{5}} \cdots\right)$$

i.e. the pair-wise interaction energy multiplied by a dimensionless, geometrical term characteristic of the structure. The latter is by no means easy to evaluate because it is a series which converges very slowly, presenting a considerable problem in the mathemetics of three-dimensional geometry.[18] It turns out, however, that the electrostatic energy per mole of electrovalent crystal of any valency type is $-NMZ_+Z_-e^2/r_0$, where N is Avogadro's number and M, the *Madelung constant* is the summation of the geometric terms. It is knowable to any degree of accuracy; some values for common structures are: rock-salt, 1·74756; caesium chloride, 1·76267; fluorite, 5·03878.

It is clear that, in this treatment, the ions have really been considered as point charges, situated at a distance r_0 apart. They are, of course, finite in size, and r_0 is a function of the balance between forces of attraction and of close-range, quantum-mechanical repulsion. To a first approximation, the latter can be expressed by B/r^n, where B is a constant, and the exponent, n, is large, varying between 5 and 13. Taking this repulsion energy into account, the lattice energy becomes, treating interionic distance as a variable,

$$U_0 = \frac{NMZ_+Z_-e^2}{r} - \frac{B}{r} \tag{3.24}$$

When $r = r_0$, the equilibrium distance of minimum energy, $dU_0/dr = 0$, so that B can be eliminated, giving

$$U_0 = \frac{NMZ_+Z_-e^2}{r_0}\left(1 - \frac{1}{n}\right) \tag{3.25}$$

which is the Born–Landé equation. X-ray crystallographic data provide r_0, and n can be calculated from compressibility,

$$\beta = \frac{1}{V}\left(\frac{\partial V}{\partial P}\right)_T,$$

as, without detail, can be seen intuitively.

The later Born–Mayer equation

$$U_0 = \frac{NMZ_+Z_-e^2}{r_0}\left(1 - \frac{\rho}{r_0}\right) \tag{3.26}$$

using an exponential dependence of repulsion on distance, is now preferred; ρ is a constant, also evaluated from compressibility, which turns out to vary little from 0·345 (r_0 in Å) from one crystal to another. Further refinements— allowance for interionic dispersion forces, the zero-point energy of crystal vibration, multipole interactions between polyatomic ions[18]—need not concern us. The main interest will be, how well do 'cycle' (Born–Haber) and 'calculated' (extended Born–Mayer) values of U_0 agree with each other?

Before answering this, it is desirable to consider how such a comparison can be fairly made. It is to be remembered that the older, clear-cut distinction between electrovalency and covalency has become somewhat blurred— they now stand at the extremes of a range of chemical interactions. It is, for example, possible to set up sequences of crystalline substances to demonstrate, in terms of structures, progressive trends from electrovalency towards covalency, or from ionic towards metallic bonding. It is therefore essential to choose for the comparison in the first instance only crystals most likely on other criteria, to be purely ionic—if such indeed exist. The prerequisites will be, ions of noble gas structure disposed in the crystal lattice in a manner determined only by ionic radius ratio, the minimisation of electrostatic energy, and the requirement of electrical neutrality. Each ion must be surrounded by equivalent, equidistant nearest neighbours of opposite sign of charge. Macroscopic properties to be looked for are familiar: hardness, involatility, near-zero electrical conductivity in the solid state, high conductivity in the fused state.

Yet another limitation is the scarcity of independently measured electron affinities or standard enthalpies of formation of gaseous anions (cf. Fig. 3·7; $\frac{1}{2}D(X_2) - E_a$) to use in the Born–Haber cycle calculations. The net result is that good agreement in the proposed comparisons is hardly to be expected for much else than the alkali or alkaline earth halides.

Within this limited field, the agreement between 'cycle' and 'calculated' lattice energies is satisfactory to the nearest few kcal, as illustrated by the sample entries in Table 3.8. This being so, it is reasonable tentatively to

Table 3.8   Comparison of Born–Haber cycle and calculated lattice energies, U_0, kcal mole^{-1}

	Born–Haber	Calc.	Diff.		Born–Haber	Calc.	Diff.
LiF	244	247	—	MgF$_2$	695	696	—
NaF	216	219	—	MgI$_2$	553	474	79
NaCl	185	186	—	CaF$_2$	619	624	—
KCl	168	169	—	CaI$_2$	494	453	41
KI	155	153	—	BaF$_2$	553	560	—
CsF	175	179	—	BaI$_2$	447	438	9
CsI	146	144	—	MnF$_2$	663	660	—
				MnI$_2$	569	480	89
AgF	228	221	7	CdF$_2$	663	632	31
AgCl	216	199	17	CdI$_2$	582	475	107
AgBr	212	193	19	PbF$_2$	596	582	14
AgI	212	185	27	PbI$_2$	515	458	57

adopt such agreement as a further criterion of purely ionic bonding, and to see whether the substantial deviations found in other cases can be systematised, and used to diagnose the incidence of other types of bonding. Favourably to this objective, it is found that when there is significant disagreement, the Born–Haber lattice energy is always the greater of the two, consistent with an enhancement of crystal stability attributable to some superimposed, non-ionic bonding. The silver halides (Table 3.8), for example, show increasing discrepancies from fluoride to iodide. That this is to be expected can be shown by comparing silver and rubidium halides. It has already been noted that the nuclear charge of $Ag^+([Kr]4d^{10})$ is less well screened than that of $Rb^+([Kr])$. The silver ion has the smaller radius $(1 \cdot 13$ Å, $1 \cdot 48$ Å$)$ and the higher electron affinity $(=$ first ionisation potential; 175, 96 kcal mole$^{-1})$, and the greater power to *deform* its anionic partner. The halide ions in sequence from fluoride to iodide have increasing radii $(1 \cdot 38, 1 \cdot 82, 1 \cdot 98, 2 \cdot 20$ Å$)$ and increasing deformabilities (assessed by polarisabilities, measured optically; $0 \cdot 96, 3.60, 5 \cdot 00, 7 \cdot 60$ cm$^3 \times 10^{24})$; they are increasingly vulnerable to Ag^+, but not to Rb^+. The silver salts, but not the rubidium salts, have a marked trend towards covalency in the expected order.* Silver iodide is whole-heartedly abnormal. Whereas silver fluoride, chloride and bromide have the rock-salt structure, silver iodide is polymorphic, with zinc blende or wurtzite structures, but at 146 °C, the silver lattice melts, and Ag^+ ions become mobile in an iodine lattice, conferring high ionic conductivity. It is a general rule that lattice energy discrepancies are accompanied by failure of the other criteria for ionic bonding, and also by loss of dependence of 'interionic' separation on a self-consistent set of ionic radii.

The divergence between cycle and calculated lattice energies becomes greater with increasing cationic charge, as might be expected. Entries in Table 3.8 illustrate this, together with the influence of cation radius and departure from noble gas structure. For the cases shown, there is a change of structure on passing from fluoride to iodide of the same bivalent metal. The change is from a structure to be expected from the most symmetrical packing of ions suitable to minimise electrostatic energy to a layer lattice of less than maximum symmetry—with an element of stereochemistry typical of covalent interaction. For transitional metal ions with unfilled d-shells, there is a splitting of the otherwise degenerate d-orbitals, and a crystal field stabilisation effect, reflected, for example, in a double maximum in U_0 along the series calcium to zinc, being zero for $3d^0$, $3d^5$ and $3d^{10}$.

Such applications of lattice energies, and many others, are sometimes hampered by lack of knowledge of the crystal structure, essential to derive M and r_0 for use in calculating U_0 by way of equation (3.26). This restriction has been partially eliminated by Kapustinskii,[19] who observed that M/vr_0, where Nv is number of ions per mole, is constant to a surprisingly close approximation for a wide range of different structures. He therefore proposed to 'reconstruct imaginatively' all other structures into an 'isoenergetic' rock-salt structure, for which $M/v = 0 \cdot 874$, and the 6-co-ordinate ionic radii are well established. Then, by assigning an average value of 9 to n, the

* Fajans's rules (1923) seem to be less well known to students than formerly. They are that transition from electrovalency to covalency is favoured by high ionic charges, small cation of non-noble gas structure, large anion.

repulsion exponent in equation (3.25), or putting $\rho = 0.345$ in equation (3.26), he arrived at the corresponding expressions

$$U_0 = \frac{256vZ_+Z_-}{r_+ + r_-} \tag{3.27}$$

and $$U_0 = \frac{287.2vZ_+Z_-}{r_+ + r_-}\left\{1 - \frac{0.345}{r_+ + r_-}\right\} \tag{3.28}$$

Despite the empiricism involved, these equations have proved to be serviceable within their limitations; to quote an authoritative review,[18] they remain 'a useful guide to approximate lattice energies, especially in those cases where the structure is unknown'. They have a positive advantage in containing separate 6-co-ordinate ionic radii, since they can be used under suitable conditions to determine effective 'thermochemical radii' of complex anions, NH_2^-, IO_3^-, ClO_4^-, PO_4^{3-} and the like. Thus, reference to the Born–Haber cycle (Fig. 3.7) shows how the difference between the lattice energies of two compounds, M_1X, M_2X with a common anion can be derived from accessible thermal data. Then, if the cation radii are known, two simultaneous Kapustinskii equations can be solved for the required anion radius. If the latter is not to be assigned overly much physical significance, it is at least a useful parameter for the calculation of the lattice energies of other crystals in which it occurs.

3.11   Some further applications

It would not be defensible to omit some examples of the powers of lattice energies, incorporated in the Born–Haber cycle, to illuminate many aspects of chemistry. Some examples accordingly follow.

3.11.1   Standard enthalpies

The calculation of standard enthalpies of formation of gaseous anions, and, where appropriate, derivation from them of electron affinities. Table 3.9 is taken from Waddington's review.[18]

Table 3.9   Some standard enthalpies of formation of gaseous anions, kcal mole^{-1}

O^{2-}	217	NH_2^-	15	C^{2-}	−245
S^{2-}	152	NH^{2-}	261	N_3^-	35
Se^{2-}	165	CN^-	7	HF_2^-	150
Te^{2-}	145	BH_4^-	−23	CNO^-	−63
OH^-	50	O_2^-	−19	NO_3^-	−81
SH^-	−31	O_2^{2-}	171	BF_4^-	−406

3.11.2   Proton affinities

The cycle of Fig. 3.8 was first used by Grimm in 1927 to derive the proton affinity of ammonia, P_{NH_3}, which is seen to be equal to

$$P_{NH_3} = \Delta H_f^\circ[NH_3(g)] + \Delta H_f^\circ[H^+(g)] + \Delta H_f^\circ[X^-(g)] - \\ - \Delta H_f^\circ[NH_4X(s)] - U$$

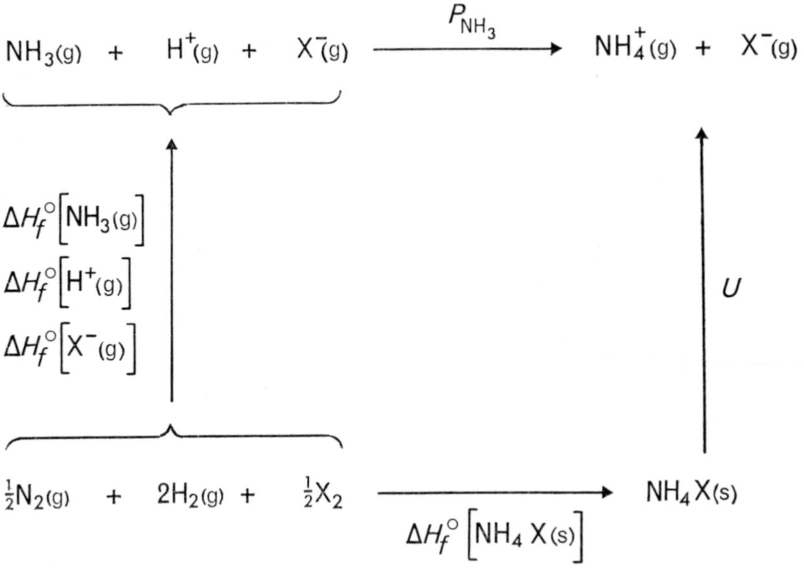

Fig. 3.8   Cycle to derive the proton affinity of ammonia

Putting in figures for three ammonium halides:

	NH₄Cl	NH₄Br	NH₄I
$\Delta H_f^\circ[NH_3(g)]$	−11	−11	−11
$\Delta H_f^\circ[H^+(g)]$	367	367	367
$\Delta H_f^\circ[X^-(g)]$	−59	−56	−47
$\Delta H_f^\circ[NH_4X(s)]$	75	65	48
−U	−156	−151	−142
P_{NH_3}	216	214	215 kcal

The proton affinity of ammonia (its true, 'in vacuum', Brønsted basic strength) is large. A result of 216 kcal mole⁻¹ has been obtained by a method based on the scattering of protons in gaseous ammonia.[11] It is of interest that ammonium fluoride gives an inconsistent result. This is because it has an open, tetrahedrally co-ordinated structure similar to that of ice (in which it is soluble), due to the presence of N—H $\cdots$ F hydrogen bonds, and these augment the lattice energy by about 15 kcal mole^{-1}. The proton affinity of the water molecule has been determined as 182 kcal mole^{-1} by assuming the same lattice energy for the isomorphous $H_3O^+ClO_4^-$ and $NH_4^+ClO_4^-$. Some other data of interest[18] are appended to Table 3·10.

Table 3.10   **Proton affinities, kcal mole^{-1} ***

NH$_3$	209†	NH$_2^-$	393	NH^{2-}	613
H$_2$O	182	OH$^-$	375	O^{2-}	554
		SH$^-$	342	S^{2-}	550

* These data come from ref. 18.
† The best accepted value is 214 kcal mole^{-1} recorded by Ladd and Lee.

3.11.3   The study of non-existent compounds

The structure a hypothetical compound would have can be credibly assumed to be the same as that of a known compound of an element of the nearest atomic number and the same stoichiometry. The lattice energy can then be assessed, and the Born–Haber cycle used to see why the compound does not exist, or even whether it is worth looking for. To take an example, consider that I_1 for calcium (141 kcal mole^{-1}) is far lower than $I_1 + I_2$ (415 kcal mole^{-1}); is it then conceivable that the monochloride, CaCl could exist? This can be studied by assigning it the same structure and lattice energy as KCl, and by making comparison with the stable compound, CaCl$_2$, in the following way:

CaCl (kcal mole^{-1})		CaCl$_2$ (kcal mole^{-1})	
Ca(s) → Ca(g)	42	Ca(s) → Ca(g)	42
Ca(g) → Ca$^+$(g) + e$^-$	141	Ca(g) → Ca^{2+}(g) + 2e$^-$	415
$\frac{1}{2}$Cl$_2$(g) → Cl(g)	29	Cl$_2$(g) → 2Cl(g)	58
Cl(g) + e$^-$ → Cl$^-$(g)	−87	2Cl(g) + 2e$^-$ → 2Cl$^-$(g)	−174
$-U$[CaCl] ($\sim U$[KCl])	−169	$-U$[CaCl$_2$]	−532
Ca(s) + $\frac{1}{2}$Cl$_2$(g) → CaCl(s) −44		Ca(s) + Cl$_2$(g) → CaCl$_2$(s) −191	

It looks, then, that CaCl might well be formed exothermally from its elements in their normal standard states. Consideration of entropy loss shows that this would not affect the issue; ΔG° for the formation of CaCl should be quite substantially negative in value. Enthusiasm is damped, however, when the two results above are combined to reveal that for $2CaCl \rightarrow Ca + CaCl_2$, $\Delta H^\circ = -103$ kcal. The compound does not exist because its disproportionation reaction has very strong thermodynamic impetus, with its origin in the high lattice energy of $CaCl_2$. There are no monohalides of any alkaline earth element for similar reasons.

3.11.4 Stabilisation of high valency by fluorine and oxygen

Since disproportionation lies in wait for higher-valent compounds, their stability, or instability, is best viewed in terms of $MX_{n+1} \rightarrow MX_n + \frac{1}{2}X_2$, for which, ignoring thermal energies,

$$\Delta H^\circ = U[MX_{n+1}] - U[MX_n] - \Delta H_f^\circ[X^-(g)] - I_{n+1}$$

We wish first to consider what happens when M is kept the same, and X is changed through the sequence of halogens from fluorine to iodine. We note that $\Delta H_f^\circ[X^-(g)] = \frac{1}{2}D[X_2] - E_a$ has the values -65, -59, -56, -47 kcal mole^{-1}, and that this itself puts the halogens in the order $F > Cl > Br > I$ in tendency to stabilise the higher oxidation state. It is, however, the lattice energies that are more significant, and to see how, it is useful to look at Kapustinskii's simpler equation, namely,

$$U = \frac{256vZ_+Z_-}{r_+ + r_-} \tag{3.27}$$

For MX_{n+1}, v (no. of ions per 'molecule') $= (n + 2)$; $Z_+Z_- = (n + 1)$; for MX_n, $v = (n + 1)$; $Z_+Z_- = n$. Ignoring for the moment the difference between the radii of $M^{(n+1)+}$ and M^{n+} (the former will be the smaller), the change in lattice energy accompanying disproportionation can be written

$$\Delta U = \frac{256}{r_+ + r_-}\left\{(n + 2)(n + 1) - (n + 1)n\right\}$$

$$= \frac{512}{r_+ + r_-}(n + 2)$$

and is clearly positive; this tends to make ΔH° the more positive and disfavours decomposition. The ignored cation radius difference would add a positive increment to ΔH°.

Consider next the influence of r_-, the anion radius. The smaller it is, the greater the positive value of ΔU. The halide ion radii (1·33, 1·82, 1·98, 2·20 Å) also establish the stabilising sequence $F > Cl > Br > I$. It can also be seen that this effect will become proportionally less as r_+ increases along a sequence of cations, thus 'bulking up' the denominator, $r_+ + r_-$; this is why the difference between the lattice energies of CsF and CsI ($179 - 144 = 35$ kcal mole^{-1}) is less than that between LiF and LiI ($247 - 177 = 70$ kcal mole^{-1}).

An interesting repercussion of the effect of cation radius on the difference of lattice energies of saline fluorides and higher halides is to be seen in organic fluorination reactions, $R - Cl + MF = R - F + MCl$, for which

$$\Delta H^\circ = -U[MCl] + U[MF] + \Delta H_f^\circ[R - F] - \Delta H_f^\circ[R - Cl]$$

The generally higher lattice energy of MF than MCl is a disability, but this is minimised by using a fluoride of a large cation. Hence CsF is a more effective fluorinating agent than LiF. If any non-ionic bonding enhances the lattice energies increasingly on passing from fluoride to higher halide, this is a great help—this is why AgF is one of the most effective fluorinating agents, and HgF_2 even better.

Oxygen also stabilises high valency; the radius of O^{2-}, $1\cdot42$ Å, and its double charge are favourable to high lattice energies, but it has a disability in its very high standard enthalpy of formation, ca. 220 kcal mole^{-1}. High-valent oxides are consequently less stable than the corresponding high-valent fluorides, although there are familiar examples of both, e.g. fluorides and oxides of Ag(II), Mn(IV), Co(III), Pb(IV), Ce(IV).

Although these calculations provide convincing explanations of well-known chemical facts, it must not be supposed that they are anything like complete. Thus, fluorine and oxygen promote the stability of high oxidation states in compounds (or complex anions) which cannot be considered electrovalent, as witness the volatile RuO_4 and OsO_4, but discussion cannot be extended.

3.11.5   Hydration and ligation energies

It is, of course, impracticable to deal with these topics; they would open up new fields—ones, moreover, in which 'first law thermodynamics' would no longer be an adequate approximation. Enough has been done for the current purpose, and the reader is left to wonder whether the second law can perhaps be as fruitful in chemical application as the first law?

REFERENCES

1. Bragg, W. L., and Williams, E. J., *Proc. Roy. Soc.*, 1934, **A, 145**, 699.
2. Eucken, A., and Veith, H., *Z. phys. Chem.*, 1938, **B, 38**, 393.
3. Moelwyn-Hughes, E. A., *Physical Chemistry*, Pergamon, 1957, p. 90.
4. NBS, Circular 500 and later issues.

5. Argue, G. R., Mercer, E. E., and Cobble, J. W., *J. Phys. Chem.*, 1961, **65,** 2041.
6. Cottrell, T. L., *J. Chem. Soc.,* 1948, 1448.
7. Cottrell, T. L., *The Strength of Chemical Bonds*, 2nd ed., Butterworths, 1968, p. 104.
8. Cotton, F. A., and Wilkinson, G., *Advanced Inorganic Chemistry*, 2nd ed., Interscience, 1966, p. 98.
9. Skinner, H. A., and Pilcher, G., *Quart. Rev.*, 1963, **17,** 264.
10. Janz, G. J., *Quart. Rev.*, 1955, **9,** 229; *Thermodynamic Properties of Organic Compounds*, revised ed., Academic Press, 1967.
11. Vedeneyev, V. I., Gurvich, L. V., Kondrat'yev, V. N., Medvenev, V. A., and Frankevich, Ye. L., *Bond Energies, Ionisation Potentials and Electron Affinities*, Arnold, 1966.
12. See ref. 8 above, p. 100.
13. Pauling, L., *The Nature of the Chemical Bond*, 3rd ed., Cornell, 1960.
14. Allred, A. L., and Rochow, E. G., *J. Inorg. Nuclear Chem.*, 1958, **5,** 264; Allred, A. L., *J. Inorg. Nuclear Chem.*, 1961, **12,** 215.
15. Allred, A. L., and Hensley, A. L., *J. Inorg. Nuclear Chem.*, 1961, **17,** 43.
16. Bartlett, N., *Proc. Chem. Soc.*, 1962, 218; Bartlett, N., and Lohmann, D. H., *Proc. Chem. Soc.*, 1962, 115; *J. Chem. Soc.*, 1962, 5253.
17. Johnson, D. A., *Some Thermodynamic Aspects of Inorganic Chemistry*, Cambridge, 1968, p. 35.
18. Waddington, T. C., in Emeléus, H. J., and Sharpe, A. G., *Advances in Inorganic Chemistry and Radiochemistry*, Academic Press, 1959, **1,** 157.
19. Kapustinskii, A. F., *Quart. Rev.*, 1956, **10,** 283.
20. Sherman, J., *Chem. Rev.*, 1932, **11,** 93.
21. Dasent, W. E., *Non-existent Compounds*, Arnold, London, 1965.

4

Mainly on Work and Entropy

4.1   Introduction

This chapter, like the last, needs a word on the what and why of its contents. It begins on conventional lines and proceeds (with a mathematical digression) normally to deal in formal terms with free energy and entropy. The main effort is to reinforce appreciation of the latter concept. The third law brings a difficulty in that its bald statement raises many questions, and to leave them unanswered might shake the reader's confidence in the writer's honesty of purpose. So challenged, the writer has done his best to deal simply with topics generally classified as advanced, and rather difficult because the mental models we all make are inappropriate. In this, he has had to assume wider background knowledge in other fields (e.g. basic wave mechanics) than in thermodynamics. This is not out of line with the general thesis of the book, but readers who find difficulties because they lack the assumed background can skip or defer without undue damage. At least the closing section should be accessible enough.

4.2   Internal energy and enthalpy

The first law of thermodynamics is often expressed by the equation

$$\Delta E = q - w \tag{4.1}$$

where ΔE is the change in the *internal energy* of a system consequent on any finite process it undergoes, q is the heat absorbed by the system from its surroundings, and w is the work done by the system on its surroundings during the process—the *external work*. The internal energy, E, is an extensive property of a system. For any process, the change in internal energy depends only on the final and initial states of the system concerned, and is independent of the path between them, i.e.

$$\Delta E = E_2 - E_1 \tag{4.2}$$

where E_2 and E_1 are the respective internal energies of the system in its final and initial states.

It is otherwise for q and w. Each is *dependent on path*, although the difference between them *is not*. The external work, w, includes any kind of energy other than heat lost by the system to the surroundings in the course of the process. Equation (4.1) therefore goes beyond the first law in singling out heat as unique among forms of energy.

If the process occurs under constant external pressure and there is no external work except that of volume change, ΔV, at the prevailing constant pressure, P.

$$\Delta E = E_2 - E_1 = q_P - P\Delta V = q_P - P(V_2 - V_1) \tag{4.3}$$

where q_P denotes heat absorbed at constant pressure. This symbol is reserved for reactions or processes occurring 'unharnessed', irreversibly, as in an open calorimeter. For any defined process, on which the condition of constancy of pressure has been imposed, $P\Delta V$ is clearly to be classed as 'obligatory work'. If, in addition, there is external work of some other kind (e.g. electrical work extracted from a reaction taking place in an electrochemical cell), ΔE is unaffected—it is independent of the way the process is conducted—but the heat absorbed is reduced by an amount equal to the additional work, and must no longer be denoted q_P. Sometimes additional state-controlling variables (e.g. surface area, gravitational, electric or magnetic fields), may be significant in defining initial and final states, giving rise to other kinds of obligatory work. Excluding such special circumstances (in which the thermodynamic treatment can be appropriately modified), equation (4.3) is applicable to all processes in closed systems.

Re-arrangement of this equation gives

$$q_P = (E_2 + PV_2) - (E_1 + PV_1) \tag{4.4}$$

where q_P appears as a difference between two terms, like in form, which are identifiable as extensive properties of final and initial states of the system concerned under the constant pressure P. These are recognised as the enthalpies, and the formal definition of enthalpy is seen to be

$$H = E + PV \tag{4.5}$$

If a process is constrained to occur at constant volume. ΔV is zero, and the external work is zero, so that

$$\Delta E = q_V \tag{4.6}$$

the heat absorbed at constant volume.

To express the dependence of internal energy, E, on temperature, use must be made of the heat capacity at constant volume, namely

$$C_V = \left(\frac{\partial E}{\partial T}\right)_V \tag{4.7}$$

which, like C_P, is an extensive property of a system.

It is desirable to give some thought to the relations between these constant pressure and constant volume functions.

For a process at constant pressure, ΔH differs from ΔE by the external work, $P\Delta V$, i.e.

$$\Delta H = \Delta E + P\Delta V, \quad \text{or} \quad q_P = q_V + P\Delta V \qquad (4.8)$$

If, for instance, an exothermic reaction normally accompanied by expansion at constant pressure (e.g. $Zn + H_2SO_2(aq) = ZnSO_4(aq) + H_2(g)$) were to be carried out in a closed calorimeter, the total volume would be kept constant, and ΔV would be zero. More heat would be evolved under these conditions than for the same reaction allowed to occur in an open calorimeter, because no energy would be consumed in the performance of external work. In the particular example quoted, q_V would be more negative than q_P by very nearly RT cal because ΔV (at constant P) is mainly due to the generation of one mole of gas. If it is desired to express determinable ΔH and ΔE values in terms of differences between H and E values—characteristic but not absolutely determinable extensive properties of final and initial states—then equation (4.5) gives a definition of H relative to E that correctly synthesises all the observable differences between ΔH and ΔE for finite processes. Physical significance can be attached to the PV term in equation (4.5), for a system not undergoing change, by imagining the hypothetical creation of the system in the ambience of its surroundings at pressure P. Room must be made for it; a cavity of volume V must be generated in the surroundings into which the system may be placed. Energy equal to the product PV must be expended to perform this displacement of the surroundings, and must be accounted for in assessing the total energy of the system in its stated equilibrium condition.

Cursory inspection of equation (4.5) might leave the impression that the internal energy, E, does not itself depend on P or V. This would be true only for the hypothetical ideal gas, but is wrong in principle for any other kind of system. How this is can best be illustrated under the next sub-heading.

4.3   Heat capacities at constant pressure and constant volume

The *general* relation between C_P and C_V for any kind of system must be argued out in mathematical language. This will afford an illustration, as previously promised, of how such argument can lead to a possibly unforeseen result of important and demonstrable physical significance. A basic preamble cannot be omitted.

Emphasis has already been laid on the significance of an extensive property, X, as a function of the state of a system, such that $\Delta X = X_2 - X_1$ for a change from state 1 to state 2, independently of path. An equivalent definition is that for a cyclic process (returning a system after a sequence of changes precisely to its initial state)

$$\oint dX = 0 \tag{4.9}$$

Yet a third equivalent statement is that dX is an *exact, complete,* or *perfect differential.* This means that

$$\int_1^2 dX = X_2 - X_1 \tag{4.10}$$

i.e. $\displaystyle\int_1^2 dX$ integrates to a finite difference independent of the path of integration. It is suitable to illustrate the properties of a perfect differential in terms of $X = E$, the internal energy of a given system. Choosing temperature and volume as independent variables of state, we can write $E = f(T,V)$, or, in frequently used notation

$$E = E(T,V) \tag{4.11}$$

meaning that E is a function of T and V. If we quote specific values for the T and V of the defined system, its E (in company with all its other properties) is uniquely fixed by the system itself and is not subject to adjustment. The differential dependence of E on T and V can be expressed in terms of dE as a perfect differential by

$$dE = \left(\frac{\partial E}{\partial T}\right)_V dT + \left(\frac{\partial E}{\partial V}\right)_T dV \tag{4.12}$$

which, so written, is called a *total differential.* Expressed for clarity in the simpler notation

$$dE = M dT + N dV \tag{4.13}$$

it can easily be remembered that

$$\left(\frac{\partial M}{\partial V}\right)_T = \left(\frac{\partial N}{\partial T}\right)_V \tag{4.14}$$

or, in terms of equation (4.12)

$$\left\{\frac{\partial}{\partial V}\left(\frac{\partial E}{\partial T}\right)_V\right\}_T = \left\{\frac{\partial}{\partial T}\left(\frac{\partial E}{\partial V}\right)_T\right\}_V \qquad (4.15)$$

or

$$\frac{\partial^2 E}{\partial V \partial T} = \frac{\partial^2 E}{\partial T \partial V} \qquad (4.16)$$

which indicates that it is immaterial in which order the differentiations are carried out. This is Euler's criterion of a perfect differential, or the Euler reciprocity relation. The act of writing down equation (4.14) after inspection of equation (4.13)—or (4.15) from (4.12)—is called cross-differentiation. This is a basic, trouble-saving operation of immense value.

Reverting to the purpose of studying the relation between C_P and C_V for any kind of system, a start can be made with equation (4.12), which is quite general. Imposing the condition of constant pressure, *divide* by dT and so obtain

$$\left(\frac{\partial E}{\partial T}\right)_P = \left(\frac{\partial E}{\partial T}\right)_V + \left(\frac{\partial E}{\partial V}\right)_T\left(\frac{\partial V}{\partial T}\right)_P$$

But, from equation (4.5), $E = H - PV$, so that

$$\left(\frac{\partial E}{\partial T}\right)_P = \left(\frac{\partial H}{\partial T}\right)_P - P\left(\frac{\partial V}{\partial T}\right)_P$$

Hence,

$$\underset{\underset{C_P}{\uparrow}}{\left(\frac{\partial H}{\partial T}\right)_P - P\left(\frac{\partial V}{\partial T}\right)_P} = \underset{\underset{C_V}{\uparrow}}{\left(\frac{\partial E}{\partial T}\right)_V} + \left(\frac{\partial E}{\partial V}\right)_T\left(\frac{\partial V}{\partial T}\right)_P$$

Therefore

$$C_P - C_V = \left(\frac{\partial V}{\partial T}\right)_P\left[P + \left(\frac{\partial E}{\partial V}\right)_T\right] \qquad (4.17)$$

a general relationship with the following translation.

C_P exceeds C_V because energy is consumed in work of expansion—against external pressure P and against *internal pressure* (not externally detectable), $(\partial E/\partial V)_T$. The latter is a function of any attractive forces acting within the system, such as lend cohesion to condensed phases. The only system with no such forces is the ideal gas, with internal energy depending only on temperature. In this case, $(\partial E/\partial V)_T = 0$, and (per mole) $V = RT/P$; $(\partial V/\partial T)_P = R/P$, so that $C_P^{\circ} - C_V^{\circ} = R = 1.987$ cal $°K^{-1}$ mole^{-1}. Real gases, not too near the critical region, follow this behaviour quite closely because they have much the same large coefficient of expansion and small $(\partial E/\partial V)_T$.

The situation is different for condensed states of matter, since $(\partial E/\partial V)_T$ is, clearly, likely to be large. It is perhaps best illustrated by a development of equation (4.17)—to be presented shortly. It is

$$C_P - C_V = \alpha^2 VT/\beta \qquad (4.18)$$

where α and β are coefficients of expansion and of compressibility,

$$\frac{1}{V}\left(\frac{\partial V}{\partial T}\right)_P \quad \text{and} \quad -\frac{1}{V}\left(\frac{\partial V}{\partial P}\right)_T$$

Solids, compared with gases, have smaller molar volume, V, and very variable values of α, smaller by one to several orders of magnitude. This is the dominant factor, and $C_P - C_V$ is small to negligible, depending on the nature of the solid. Liquids are not intermediate in behaviour between solids and gases, and a general reason why can be found in our recurrent theme of order–disorder, energy–entropy balance. For solids and gases, this balance lies on one side or the other, and each shows its own degree of regularity of behaviour in this respect. It is in the liquid phase that the battle is joined. Coefficients of expansion are greater than for solids; typically, molar volume is trebled between triple and critical points. The internal work of expansion may become very large, with a corresponding large difference between C_P and C_V, strongly dependent on temperature. However specific and peculiar the behaviour may be (it is illustrated in Table 4.1) it remains in accordance with equation (4.18) and explicable in terms of equation (4.17)—as far as it goes.

Table 4.1   Heat capacities at constant pressure and at constant volume, cal $°K^{-1}$ mole^{-1} or cal $°K^{-1}$ (g atom)$^{-1}$

Substance	Temp., °K	C_P	C_V	$C_P - C_V$
Diamond	293	1·442	1·441	0·001
Sulphur (α)	293	5·64	5·21	0·43
Aluminium	298	5·82	5·59	0·23
Argon	85	10·0	4·7	5·3
Oxygen	85	12·9	7·0	5·9
Benzene	293	32·2	22·6	9·6
Diethyl ether	293	41·1	30·1	11·0
Water	273	18·18	18·11	0·07
Water	373	18·12	15·91	2·21

Some general comments can usefully be made. C_P, even measured at $P = 0$, is not, in principle, to be equated to C_V. Nobody sets out to measure C_V by a *direct* experimental method (almost impossibly difficult) because it can be calculated from the other, more easily measured, quantities in equation (4.18)—thermodynamics appears in the guise of a labour-saving device. While C_P is the experimental quantity, theoretical work on heat capacity is usually concerned with C_V—perhaps to be understood in the light of earlier discussion. Finally, it becomes apparent why, for most purposes, P and T form the more judicious choice than V and T as independent variables of state.

4.4 Maximum work and reversible processes

Any spontaneous process can, in principle, be harnessed to provide a quantity of useful work. The question arises, are there any general principles that must be observed in order to obtain the maximum possible work from any given process? Obviously, there must be an upper limit in each case.

It may seem a little odd that this problem is traditionally (and very usefully) explored in terms of an atypical process taking place neither at constant pressure, nor at constant volume. It is the isothermal (constant temperature) expansion of one mole of ideal gas.

Imagine the gas to be confined by means of a piston in a cylinder in thermal contact with a heat source, or heat reservoir, which will maintain it at a constant temperature, T. The piston can be used to apply an external pressure, P_{ext}, opposing the pressure P, exerted by the gas itself. If $P_{ext} = P$, there is a state of equilibrium, and nothing happens. If the gas is to expand, P_{ext} must be less than P, so that the piston moves—providing the mechanism by which the work is obtained. The work is the product of the expansion and the pressure against which it occurs. In general, whether P_{ext} is constant or not,

$$ w = \int_{V_1}^{V_2} P_{ext}\, dV $$

for a finite expansion between the volume limits V_1 and V_2.

To maximise the work obtained from this expansion P_{ext} must be as large as possible throughout. But P_{ext} must be at least infinitesimally smaller than P, otherwise expansion will no longer be a natural process and will not occur spontaneously. Hence, the maximum work will be

$$ w_{max} = \int_{V_1}^{V_2} (P - dP)\, dV $$

In the limit, of course, the second-order infinitesimal $dPdV$ vanishes, so that, without approximation,

$$w_{max} = \int_{V_1}^{V_2} PdV \tag{4.19}$$

where P is the pressure *of the gas itself*. Since this is known as a function of the volume for the system concerned $(P = RT/V)$, the integration can be performed to give

$$w_{max} = RT \ln (V_2/V_1) = RT \ln (P_1/P_2) \tag{4.20}$$

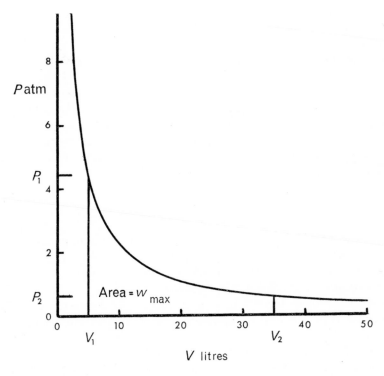

Fig. 4.1 Graphical representation of the reversible, isothermal expansion of one mole of ideal gas

The graphical version of this integration is shown in Fig. 4.1, for 1 mole of ideal gas expanded from 5 to 35 litres at 0 °C.*

It is apparent from equation (4.20) that w_{max} is determined only by the final and initial states of the system, and therefore has the status of a change in an extensive property of the system. It was reasonable to call this property the *work content* (symbol A for Arbeit), so that the maximum work got out of the system in the course of the process is equated to its loss in work content:

$$w_{max} = -\Delta A = -(A_2 - A_1) \tag{4.21}$$

A process conducted in this way is said to be a *reversible process*. This is because, for example, the gas will expand spontaneously when $P_{ext} = P - dP$, but will contract if P_{ext} is raised to $P + dP$. Thus, an infinitesimal change in the external condition, P_{ext}, will reverse the direction of the natural process. Furthermore, the maximum work obtainable from the process is equal to the minimum work required to reverse it.

Reversibility is the condition essential to obtaining the maximum work from any spontaneous process. The process must be opposed by a 'work-harnessing load' as great as will allow it to remain spontaneous at all. This is the only condition to involve negligible departure (in the limit, none) from a state of equilibrium—or a developing sequence of such states evolving continuously from each other. Reversible processes must be either infinitesimal in extent, or, if finite, infinitely slow.

It is proper to ask, where does the work got from the expansion of the gas come from? This question can be examined in terms of the first law written as in equation (4.1), i.e.,

$$\Delta E = q - w \tag{4.1}$$

Whereas ΔE is independent of the path taken by the process concerned from its initial to its final state, q, heat absorbed, and w, the work done by the system on its surroundings, are not independent of path. Clearly, for the gas expansion, w can be anything between zero and w_{max} according to how the expansion is conducted—between the limits of a completely irreversible expansion (like blowing off compressed air into a vacuum) and a fully

---

* For students with no experience of graphical integration, and unaware that it is a functional operation, it is not a waste of time to make a careful plot of the Boyle's law rectangular hyperbola, conduct an integration by counting squares and compare the result with that obtained by solving equation (4.20).

reversible one. For the reversible, isothermal expansion of the mole of ideal gas,

$$\Delta E = q_{rev} - w_{max} \tag{4.22}$$

where q_{rev} is the heat absorbed by the system from its surroundings during the reversible process. For this particular process, we know that $\Delta E = 0$ (cf. Joule's classical experiment; kinetic theory evidence that the internal energy of the ideal gas depends only on T and not on V), so that

$$q_{rev} = w_{max}$$

So the work comes from the heat absorbed by the ideal gas from its surroundings as it expands reversibly at constant temperature. Therefore, from equation (4.20),

$$q_{rev} = RT \ln (P_1/P_2) \tag{4.23}$$

It may be asked, why does this happen? Why is the expansion of a gas, ideal or not, at constant temperature, a spontaneous process? The answer is to be found in the summarising statement made in chapter 2—'a system may be in disequilibrium for either or both of two reasons. It may be in a state of more than minimum energy, or it may be in a state of less than maximum entropy or probability.' The energy motive has been excluded. The gas expands because it increases in entropy by expanding. The natural process of gas expansion is exclusively 'entropy-driven'.

The extent of entropy change accompanying the reversible, isothermal expansion of one mole of ideal gas is found by rearranging equation (4.23):

$$q_{rev}/T = R \ln (P_1/P_2) \tag{4.24}$$

The quantity q_{rev}/T is seen to depend only on final and initial states. It is $\Delta S = S_2 - S_1$. The relation

$$\Delta S = q_{rev}/T \tag{4.25}$$

turns out to be the general recipe for finding ΔS for an isothermal process. Conduct the process reversibly and divide the heat absorbed by the absolute temperature.

4.5 The conversion of heat into work (an interpolation)

As just shown, this conversion is effected by the isothermal, reversible expansion of an ideal (or other) gas, heat absorbed and work obtained being equal to $T\Delta S$. But this is a fruitless, once-for-all process. To repeat it, the gas would have first to be compressed back to its initial state, and this would use up all the work, *unless* the compression were conducted at a lower temperature. In this case, the gas can be made the basis of a cycle that will repetitively supply work by harnessing the natural flow of heat from a heat source at a higher temperature, T_1, to a heat sink at a lower temperature, T_2. The simplest such cycle involves four sequential operations, all conducted reversibly, as follows.

1. A mole of ideal gas, in thermal contact with the heat source at T_1, is expanded isothermally and reversibly so that its entropy increases from S_1 to S_2. If the heat absorbed is denoted q_1, and the (maximum) work obtained w_1, then $q_1 = w_1 = T_1(S_2 - S_1)$.

2. The gas is removed from contact with the heat source, and is allowed to expand *adiabatically* and reversibly until its temperature has fallen from T_1 to T_2. Any reversible process conducted adiabatically (zero heat absorption) is (consistently with the above 'recipe') *isentropic*. So the entropy of the mole of ideal gas remains constant at S_2, and the work in this step is obtained at the expense of the internal energy—this is why the temperature falls. The loss in internal energy, and therefore the work, is, of course, solely determined by the temperature difference $T_1 - T_2$.

3. The gas is placed in thermal contact with the heat sink at T_2 and is compressed isothermally and reversibly until its entropy is restored to S_1. If, for clarity, we disregard the normal sign convention and call the heat *evolved* q_2, and the work *required* w_2, then, since the step is reversible, $q_2 = w_2 = T_2(S_2 - S_1)$.

4. The gas is removed from contact with the heat sink, and is compressed adiabatically and reversibly until its temperature is restored to T_1. The entropy remains constant at S_1, but the internal energy (a function of temperature alone) is raised back to its initial value.

The cycle is completed. Totting up, the work terms of steps (2) and (4) cancel. Heat q_1 has been absorbed at T_1, heat q_2 has been rejected at T_2. Net work, $w_{max} = w_1 - w_2$ has been obtained. From the first law, $w_{max} = q_1 - q_2$. The *efficiency* of the cycle, expressed as the fraction of the heat absorbed converted into work is seen at a glance to be

$$\frac{w_{max}}{q_1} = \frac{T_1 - T_2}{T_1} \tag{4.26}$$

which can reach its upper limit of unity only if T_2 is zero—an unattainable asymptote.

This is, of course, a version of the 'Carnot cycle'. The theorem is completed by showing that all conceivable 'reversible heat engines' working between the same two temperatures have this same efficiency. If one were more efficient than another, it could drive it backwards, and the two could operate together indefinitely, converting heat to work at a single temperature. This would contravene the second law—of which more later.

Setting aside the fundamental implications of equation (4.26) for the time being, it is of interest to mention two practical aspects.

The first is the heat pump. Suppose, given work w_{max}, we wish to supply heat at temperature T_1, to keep the house warm. We can convert it all into heat and by the first law, get heat precisely equal to w_{max}. If instead, we use the work to drive a reversible heat engine backwards, we can draw heat q_2 from outside at the lower temperature T_2 and discharge heat q_1 into the house, getting $w_{max} + q_2$ for the same money. This is the basis of schemes for communal space-heating, coming into increasing use.

The other aspect is the utilisation of fuels for energy production. In terms of the familiar relation $\Delta G = \Delta H - T\Delta S$ for the combustion, $-\Delta H$ is used to raise steam to drive turbines, etc. Nothing higher than the 'Carnot efficiency' is possible (T_1 and T_2 are steam and condenser temperatures) nor, in practice, much better than 45%. It was Ostwald in 1894 who pointed to the necessity of tapping $-\Delta G$, which is, in principle, feasible by electrochemical means. This is one reason why *fuel cells* are now receiving such active (but not yet fully rewarded) attention. The difficulties are not thermodynamic, but kinetic—the development of sufficiently active and cheap electrode catalysts.

4.6   Free energy

Reverting to the main theme, after an interpolation having virtue in emphasising that $\oint dS = 0$, we rewrite equation (4.22), i.e.

$$\Delta E = q_{rev} - w_{max} \qquad (4.22)$$

in the form

$$\Delta E = T\Delta S + \Delta A \qquad (4.27)$$

and point out with some force that this is true for whatever isothermal process concerned, whether it is conducted reversibly or not. This is so because E, S and A are extensive properties, so that ΔE, ΔS and ΔA are independent of path. If the process *is* conducted reversibly $q_{rev} = T\Delta S$ and $w_{max} = -\Delta A$. If it is conducted irreversibly, $q \neq T\Delta S$ and $w \neq -\Delta A$.

Again because E, S and A are extensive properties, an equation can be written defining A, the work content, also called the *Helmholtz free energy*, of a system. It is

$$A = E - TS \qquad (4.28)$$

Just as it was convenient to derive H from E, it is convenient to derive the Gibbs free energy, G, from A in the same way, i.e.

$$G = A + PV \qquad (4.29)$$

Bringing in equation (4.5)— $H = E + PV$ —and equation (4.28),

$$G = E + PV - TS = H - TS \qquad (4.30)$$

which defines the Gibbs free energy as an extensive property of a system.

From the definition of G relative to A in equation (4.29), it is clear that for any process conducted at constant temperature and pressure,

$$\Delta G = \Delta A + P\Delta V \qquad (4.31)$$

It is better for interpretation to rewrite this equation in terms of losses, i.e.

$-\Delta G$	=	$-\Delta A$	$-P\Delta V$
Net work obtainable from a reversible process at const. T, P		Maximum work	Obligatory work of volume change at const. press. P necessary to maintain the system in mechanical equilibrium with its surroundings.

so that the Gibbs free energy loss is a measure of the work obtainable from a reversible, isothermal process occurring at constant pressure. It is greater or less than the maximum work according to the sign of the obligatory work of volume change at constant pressure. For a process conducted at constant volume, there is no such obligatory work, so that $-\Delta G$ and $-\Delta A$ are the same. It is $-\Delta G$ which is the general measure of the affinity of a process.

For the special process discussed in detail, taking place neither at constant pressure nor at constant volume—the isothermal, reversible expansion of the ideal gas, we can write

$$dA = -PdV$$

for each infinitesimal step. General differentiation of equation (4.29) gives

$$dG = dA + PdV + VdP$$

But, in this case, $PV = $ constant, so that $PdV + VdP = 0$, and

$$dG = dA = VdP$$

For a finite expansion therefore

$$\Delta G = \int_{P_1}^{P_2} VdP \tag{4.32}$$

which leads again to equation (4.20). In this case also,

$$w_{max} = -\Delta A = -\Delta G.$$

The contribution of TS to A and G requires further discussion, and so, therefore does entropy.

4.7   Entropy

4.7.1   The dependence of entropy on temperature

Consider any system in a state of equilibrium. Let it undergo an infinitesimal process. The departure from equilibrium conditions is negligible (in the limit, zero), so the infinitesimal process is reversible. The first law therefore applies in the form

$$dE = TdS - PdV \tag{4.33}$$

where the terms on the right-hand side will be recognised as appropriate representations of q_{rev} and w_{max}.

Whether out of place or not, experience indicates the need for an emphatic comment, as follows.

In equation (4.33), and elsewhere, terms such as dE, dS and dV are infinitesimals—the limiting values of increments of E, S and V becoming indefinitely smaller and smaller. They are on no account to be confused with ΔE, ΔS and ΔV, which are finite increments. The first law for an infinitesimal change should not be written $dE = dq - dw$ because this might imply that dq and dw are perfect differentials, which they are *not*. The symbols dq and dw, and of course Δq and Δw should be rejected as absurdities.*

* If it is really necessary to represent vanishingly small quantities of heat and work, δq and δw may be used, as by Gibbs.

If the infinitesimal process suffered by the system in equilibrium is constrained to occur at constant volume, $dV = 0$, so that

$$dE = TdS \text{ at constant volume}$$

Division by dT and re-arrangement gives

$$\left(\frac{\partial S}{\partial T}\right)_V = \frac{1}{T}\left(\frac{\partial E}{\partial T}\right)_V = \frac{C_V}{T} \tag{4.34}$$

which is the general equation, applicable to any system, giving the entropy-temperature coefficient at constant volume.

Constant pressure functions are normally of greater interest. Taking the equation defining enthalpy,

$$H = E + PV \tag{4.5}$$

and differentiating generally,

$$dH = dE + PdV + VdP$$

This takes care of all possible sources of infinitesimal variation in H. But in all cases the first law must apply; substituting therefore for dE in terms of equation (4.33),

$$dH = TdS + VdP \tag{4.35}$$

If the condition of constancy of pressure be imposed on the infinitesimal process, $dP = 0$ and

$$dH = TdS \text{ at constant pressure}$$

therefore

$$\left(\frac{\partial S}{\partial T}\right)_P = \frac{1}{T}\left(\frac{\partial H}{\partial T}\right)_P = \frac{C_P}{T} \tag{4.36}$$

which, like equation (4.34) is of perfectly general application.

To find the finite increment of entropy caused by a finite rise of temperature under either imposed condition, an integration is necessary—for example,

$$\Delta S = S_{T_2} - S_{T_1} = \int_{T_1}^{T_2} \frac{C_P}{T} \, dT \tag{4.37}$$

and this requires a knowledge of C_P as a function of temperature over the range of temperature concerned.

Especial interest centres in the case where $T_1 = 0$, and this is where the third law of thermodynamics comes in.

4.7.2 The third law

This law, first specifically stated by Planck in 1912, can well be expressed essentially in the words used by Lewis and Randall in 1923[1], as follows:

Every substance has a finite, positive entropy, but at the absolute zero of temperature the entropy may become zero, and does so become in the case of a pure, perfectly crystalline substance.

The significance of this can be illustrated in two ways. First, by a quotation, now classical, from Lewis and Gibson:[2]

If the entropy of a given state be regarded in some sense a measure of the randomness of that state, the condition of a perfect crystal of a pure substance at the absolute zero is unique. In a solution there is a random distribution of several types of molecules. Even in a pure liquid or glass there is some randomness of arrangement. In any substance at a finite temperature there is a random distribution of energy among the individual molecules. But in the pure crystal at the absolute zero no randomness remains, for when the positions and properties of a few molecules are fixed, the positions and properties of all other molecules are completely determined; thus when a few elements of the crystalline structure are known we may build up the whole crystal by a process of repetition. It is this lack of any sort of randomness that we believe to be the theoretical basis for the conclusion that the entropy of a perfect crystal of a pure substance vanishes at the absolute zero.

The second illustration can be taken from what has already been written in chapter 2; randomness was clarified in terms of the impartiality of exploration by a system of all the microstates available to it. In a perfect crystal of a pure substance at the absolute zero, all molecules, atoms or ions occupy their allotted positions in the space lattice, all are in the one, lowest conceivable energy state. There is only one microstate, and the thermodynamic probability of the macrostate, W, is unity; i.e. there is only one way in which it can be realised. Hence, in terms of equation (2.2), namely, $S = k \ln W$, $S = 0$.

To extend this argument a little, consider a 'mixed crystal'—a solid solution, perfect in structure but containing equal numbers of two different kinds of atoms, A and B, randomly distributed over N available lattice sites. Even although, at zero temperature, all were in the lowest energy state, the total number of microstates would be

$$W = (N_A + N_B)!/N_A!N_B! \tag{4.38}$$

It can easily be seen, using Stirling's approximation that $\ln N! = N \ln N - N$, that this would lead to a residual entropy of $R \ln 2$ per mole, i.e. $1 \cdot 38$ cal $°K^{-1}$ $mole^{-1}$. This is why the word 'pure' is essential in the statement of the third law.

The importance of the law lies in setting up entropy as a finite, positive quantity measurable on an absolute scale. It provides the basis for the thermal, or calorimetric method of measuring standard entropies of substances. This involves measurements of C_P at $P = 0$ (or the equilibrium pressure of the system) down to very low temperatures (*ca.* 10 °K), by methods pioneered by Eucken. C_P/T is plotted against T, and the integration performed graphically, or, latterly, by more sophisticated methods. An example (the determination of the standard entropy of cadmium)[3] is shown in Fig. 4.2. The gap, 0 °K to the lowest temperature of measurement, is bridged by use of Debye's theory of the heat capacity of solids—an improvement on Einstein's earlier treatment—leading to

$$C_V = \frac{12\pi^4 R}{5}\left(\frac{T}{\theta}\right)^3 \tag{4.39}$$

valid for very low temperatures. The 'characteristic temperature', θ, for each substance is determinable from the measurements made at the lower experimental temperatures. The difference between C_P and C_V in the range concerned is small and can either be ignored, or taken into account semi-empirically with negligible error.

If phase transitions occur below $298 \cdot 15$ °K, they give rise to entropy jumps, but melting at T_M and evaporation at T_B take place reversibly under equilibrium conditions. For such transfers of material between two phases in equilibrium with each other, $\Delta G = 0$, so that $\Delta H = T\Delta S$. Hence (again in conformity with the 'recipe' previously given),

$$\Delta S = \lambda/T \tag{4.40}$$

where λ is the appropriate molar latent heat.

The method can be represented by

$$S° = S_0° + \int_0^{T_M} \frac{C_P(s)}{T} \, dT + \int_{T_M}^{T_B} \frac{C_P(l)}{T} \, dT + \int_{T_B}^{298 \cdot 15} \frac{C_P(g)}{T} \, dT$$

$$+ \frac{\lambda_M}{T} + \frac{\lambda_B}{T} \tag{4.41}$$

for a gaseous substance, with terms deleted appropriately for a liquid or a solid. The third law suggests that the term $S_0°$ may be zero. The procedure

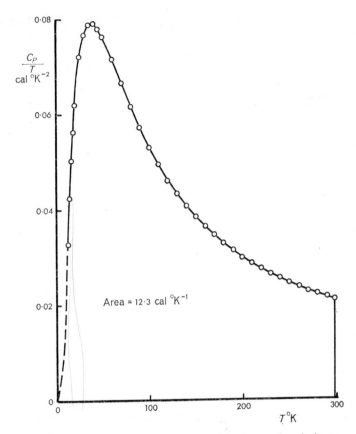

Fig. 4.2 'Third law' evaluation of the standard entropy of cadmium

for liquids is much the same as for solids. Gases, however, are usually treated alternatively. After correction to the ideal state (Berthelot's equation is used), the entropy increment from T_B to 298·15 °K is obtained, given the necessary spectroscopic data, by statistical mechanical methods—which are beyond our terms of reference.

Satisfactory support for the third law comes from the now numerous (a few hundred) examples of good agreement between $S°$ values for one and the same gaseous substance derived by alternative and independent routes. On the one hand, the value from heat capacities at very low temperatures upwards, through solid, liquid, gaseous states and the intervening transitions, based on $S_0° = 0$. On the other hand, the value obtained (without reference to solid or liquid states) by adding together contributions to entropy from the quantised molecular motions (translation, vibrations, rotations), calculable from the equations of statistical mechanics, using the data from spectroscopy. It is proper to pause a little to consider the magnitude of such agreement in terms of scientific achievement. As it stands, the methods go hand-in-hand in the continuing task of accumulating and improving the standard data.

In this situation, it is tolerably certain that when disagreement is found between the entropy results from the two methods, there is a good physical reason for it. This was so in the case of ethane, and the discrepancy was resolved when hindrance of intramolecular rotations was better understood. A more common reason is the failure of an assumption implicitly made in applying the calorimetric method. This is, that internal equilibrium is maintained down to the lowest temperatures, so that, at all stages, the energy is at a minimum. But the self-adjustments a rigid solid near absolute zero can make may be so slow as to falsify this assumption. Disequilibrium is then frozen-in, and $S_0^\circ \neq 0$. It is not quite fair to call this a failure of the third law. Carbon monoxide provides an example. The small, non-polar, compact dumb-bell molecules cannot care very much which way round they face in the crystal lattice. Crudely represented, the assembly

$$CO \quad CO \quad OC \quad CO \quad OC \quad OC \quad CO$$
$$CO \quad OC \quad CO \quad CO \quad OC \quad OC$$
$$OC \quad CO \quad CO \quad OC \quad CO \quad OC \quad OC$$

can differ little in energy from one with all the molecules facing the same way. There is little 'energy drive' towards complete order, and little 'thermal jogging'. If the two alternative orientations were equally represented and randomly distributed, the residual entropy would be $1 \cdot 38$ cal $^\circ K^{-1}$ mole^{-1}, by the argument already given. The 'spectroscopic entropy' of carbon monoxide, S_{298}°, is $47 \cdot 3$ cal $^\circ K^{-1}$ mole^{-1}. The 'calorimetric entropy' is $46 \cdot 2$ cal $^\circ K^{-1}$ mole^{-1}—*less*, because of residual disorder, leaving less order to be thermally destroyed with rising temperature. The residual entropy, $1 \cdot 1$ cal $^\circ K^{-1}$ mole^{-1} suggests that molecular orientation is not quite fully random in the crystal.

Ordinary ice (ice Ih; hexagonal) provides a specially interesting example of a well-explained entropy 'discrepancy'. In ice, each water molecule (acting twice as proton donor and twice as proton acceptor), is hydrogen-bonded to four neighbours, disposed about it in the tetrahedral directions. This, perpetuated in three dimensions, leads to the structure shown in plan and elevation in Fig. 4.3.

Each oxygen atom is surrounded by four others at an internuclear distance of $2 \cdot 75$ Å, three in the same puckered layer of six-membered rings, and the

fourth in an adjacent layer. Between the layers there are rather large poly-
hedral cavities with twelve vertices, linking up to form channels running
through the hexagonal rings. The structure is the same as that of a form of
silica—tridymite. The dielectric constant of ice is greater than that of water

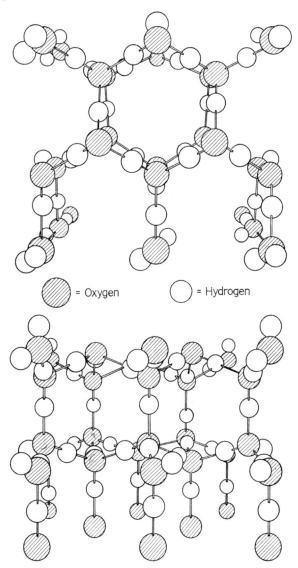

= Oxygen   = Hydrogen

Fig. 4.3 The structure of ice Ih

at 0 °C (91·2, 88·2), and proton mobility (equivalent conductance of hydrogen ions) is greater in ice than in liquid water. These electrical properties leave no doubt that the protons in ice are mobile, and cannot be orderly arranged. At very low temperatures, however, the dielectric constant of ice falls to about 4, suggesting the freezing-in of proton disorder. Adopting this assumption, consider any one oxygen atom with its four equidistant oxygen neighbours. There are four O · · · O directions away from the central oxygen, each accommodating one proton. Two of the four protons are near to the central oxygen (1·0 Å), being covalently bound to it. Two others are more distant (1·75 Å), interacting only with the lone-pair orbitals of the central oxygen atom. Since there are six ways of taking two out of four things, there are six possible arrangements of the protons, giving six alternative *orientations* to the central *water* molecule. But each of these alternatives requires two of the peripheral molecules to have favourable, as opposed to equally likely unfavourable orientations. This divides the absolute probability of any of the six arrangements of protons by four, and brings the total number of configurations (microstates) in an ice-like assembly of N water molecules to $(6/4)^N$. This would lead to a residual entropy at 0 °K of $S_0^\circ = R \ln (3/2) = 0.81$ cal °K^{-1} mole^{-1}. This is precisely the value required to bring 'third law' and spectroscopic entropies of water vapour into coincidence.[4] The disorder of protons in ice has been confirmed by neutron diffraction.

Even a brief essay on the third law, dealing as it does with 'perfectly crystalline, pure substances', raises questions it would be disingenuous to ignore, because they have fundamental implications. One of them is brought into sharp focus by the last example. The satisfactory explanation of the residual entropy of ice is all very well, but how is the proton disorder, due to proton mobility, to be explained? It *requires* the presence in ice of two kinds of *intrinsic defect*—ionic and orientational. The former comes from proton transfer $(2H_2O = H_3O^+ + OH^-)$; the latter arises by rotation of a water molecule through 120° around an O—H · · · O axis, generating empty (*leer*) and doubly proton-occupied (*doppelt*) oxygen–oxygen 'bonds' (i.e. O · · ·O and O—H · · · H—O). These are called L- and D-Bjerrum defects.[5] By rotation of adjacent water molecules they can become separated and migrate through the ice crystal. Specialised perhaps, but an example of how defects, having little *direct* effect on the state of a system in relation to its total entropy, may provide the means by which that state is established, and have a profound influence on other properties.

The general question as to the prevalence of defects in crystals can be answered by the statement that a perfect crystal is exceptional. Not only are normally encountered crystalline substances polycrystalline, consisting of randomly oriented crystal grains (a comparatively unimportant accident of sporadic nucleation of crystal growth), but they are structurally imperfect in more fundamental ways. Aside from the important class of non-stoichiometric compounds and other semi-conducting materials, it is the general rule for even pure, stoichiometric crystals to contain defects and dislocations. In ionic, or semi-ionic crystals, intrinsic defects arise in two ways. Frenkel defects (as in AgCl) are formed by displacement of an ion from a lattice site (leaving a vacancy) into an interstitial position. Schottky defects (as in NaCl) consist of a pair of vacancies of opposite sign—to be regarded as formed by removal of pairs of ions from their proper positions to the surface of the crystal. Such defects are mobile and are necessary to explain the high rates of interdiffusion and self-diffusion of solid substances. The strength of metals is reduced by orders of magnitude by *dislocations*; metals have a mosaic structure made up of internally perfect crystallites slightly disoriented with respect to their neighbours. Edge dislocations arise from a misfit between adjacent lattice planes, one of which contains an excess of atoms—the usual analogy is a ruck in a carpet. Screw dislocations are important in relation to crystal growth. Crystal surfaces are, in effect, immense dislocations, giving rise to local differences in structures and properties decisive in relation to adsorption and catalysis.

These matters, although of great interest, are not within the purview of this book—they are reviewed in an excellent standard text.[6] Nevertheless, mention of them cannot be omitted for several reasons. One is to correct misconceptions that a bald statement of the third law might engender. Another is to show that defects are not to be swept under the thermodynamic carpet; on the contrary, they illustrate thermodynamic principles in the following way.

Recalling a previous discussion of the energy–entropy balance (section 3.3), it was said that the conflict goes on all the time, and the balance shifts towards increasing entropy by any available means as the temperature rises. Every little weakness is exploited. It is now understood that whatever balance is struck at a particular temperature, it is such as to minimise the Gibbs free energy, $G = H - TS$. The formation of a defect consumes energy (dH positive), but this is paid for and more by an increase in entropy (dS positive), since a defect is a centre of disorder. In the balance, $dG = dH - TdS$, is negative. Defects are formed spontaneously and are characteristic of equilibrium states. As to how they are formed, this is a problem that requires reconsideration of the principle that thermodynamics is a science of macroscopic systems. How macroscopic does a system, or part of a system, have to be in order that the statistical averaging process will confer complete uniformity of properties? Sooner or later, with ever-decreasing size down to the molecular level, *fluctuations* must become significant. They do so well *before* the molecular level is reached. Familiar evidence for this is found in the Brownian motion of comparatively gross colloidal particles. Another aspect repaying thought is that macroscopic systems explore every micro-state consistent with constant total energy. There should therefore be no fundamental difficulty in envisaging, within the framework of thermo-dynamics, the occurrence in systems of local regions that deviate from average energy and entropy in inverse proportion to their dimensions—hot spots (and cold) that flicker into, and out of, existence entirely in accord with the principle of maximisation of probability. This provides at least a clue as to the formation of defects, as well as how, with falling temperature, they may get frozen-in, out of equilibrium, in solid systems. For systems that are in equilibrium it can also be seen that the concentration of defects will increase with rising temperature in proportion to $\exp(-E/kT)$, where E is the energy required to produce a defect, and k is the Boltzmann constant. This behaviour may be changed if defect formation becomes co-operative, with characteristics previously noted. This is not unlikely if defects are centres of weakness and become numerous. It is the general reason for the 'premonitory', pre-melting phenomena exhibited by some pure substances. The normal, calorimetric method of entropy determination of course takes care of all this; C_P may show an accelerating rise as the melting point is approached. Melting, and other phase transitions, are not as simple as the normal thermodynamic treatment (yet to come) might suggest.[7] Fluctuations (cold ones) continue to play a role in theories of the liquid state,[8] especially in the case of water, the most 'structured' of liquids.[9]

Another question likely to arise from the statement of the third law relates to the implications of the word 'pure'. If purity is essential to zero residual entropy, is no account to be taken of the fact that elements and compounds are natural mixtures of isotopically differing species? The general answer is that if species distinguishable from each other in any way are mixed, the mixture will have a non-zero residual entropy. If, however, no unmixing process need ever be contemplated, this entropy can be ignored. It *is* ignored in the *practical entropy values* used in 'run of the mill' chemical thermo-dynamics.

A final question relates to nuclear spin isomerism. This calls for a more substantial answer—given in the following section.

4.7.3   Ortho- and para-hydrogen

The story, in outline, begins with 'third law studies' of ordinary hydrogen. At room temperature, C_V is near the classical value of $5R/2$ ($R/2$ for each of 3 translational and 2 rotational 'squared terms', vibration inappreciable below 500 °K), but falls to around $3R/2$—the value for a monatomic gas—below 50 °K. This was rightly attributed to the quantal freezing-out of rotation. The moment of inertia, I, of the light H_2 molecule is small (H_2, 0.47×10^{-40}; N_2, 13.6×10^{-40} g cm^2), and the spacing of the quantised rotational energy levels correspondingly wide (energy of rigid rotator, $\varepsilon = (h^2/8\pi^2 I) J(J + 1)$, where h is Planck's constant and J is an integral quantum number, 0, 1, 2... For H_2, $\varepsilon_1 - \varepsilon_0 = 14.7 \times 10^{-3}$ eV; cf. N_2, 0.50×10^{-3} eV; kT at 40 °K, 3.45×10^{-3} eV). Calculations on this basis, however, failed to reproduce the experimental $C_V(T)$ plot.

A clue came from the rotational lines in the far uv band spectrum of hydrogen—they have a 3:1 alternation in intensity, consistent with restriction of transitions to those between rotation states of even J, 0, 2, 4 ... (with symmetrical wave functions), or between states with odd J, 1, 3, 5 ... (antisymmetrical wave functions), the latter having three times the statistical weight of the former. This was consistent with the presence of two kinds of H_2 molecule, differing in the mode of combination of the two nuclear spins.

The proton has an angular spin momentum defined by a quantum number of $i = \frac{1}{2}$ (in $h/2\pi$ units). Symmetrical (parallel) combination of two spins of $\frac{1}{2}$ gives a resultant spin of 1. This state of the molecule has 3-fold quantum weight because the quantised projection of the total spin momentum on an external axis of reference (as defined by a weak magnetic field) may be $+1$, 0, or -1. It is said to be 3-fold degenerate ($2i + 1 = 3$); it offers additional accommodation because it can be trebly filled, and is called the *ortho*-state—this name being used for which of two states has the greater statistical weight. On the other hand, antisymmetrical (antiparallel, opposed) combination of the two spins gives a resultant of zero. This state of the molecule has single weight, and is called the para-state.

If, now, symmetrical nuclear spin states combine *only* with antisymmetrical molecular rotation states, and vice versa, the spectroscopic observations are explained. The evidence therefore is that ordinary hydrogen is a mixture of three parts of orthohydrogen with one of parahydrogen, these components *separately* giving rise to rotational spectral lines, due respectively to 'J odd' and 'J even' transitions. The situation is summarised in Table 4.2.

The ortho/para ratio is, however, closely determined by the 'spin weights'

Table 4.2 Ortho- and para-hydrogen molecules

	Nuclear spin isomer	
	ortho	para
Spins	symmetrical	antisymmetrical
Resultant nuclear spin	1	0
Spin weight	3 ($+1, 0, -1$)	1 (0)
Rotation	antisymmetrical	symmetrical
J	1, 3, 5 ...	0, 2, 4 ...

only at sufficiently high temperatures (deviation from 3 : 1 is about 3%
at 0 °C), when rotational levels are plentifully populated and C_V has reached
its classical value. As the temperature falls, the *equilibrium* ortho/para ratio
decreases in a manner calculable by the equations of statistical mechanics;
it comes under control of the most probable distribution of the molecules
(of lesser total energy) over fewer, widely spaced rotational levels, with
'J even' levels forbidden to ortho-, and 'J odd' to para-molecules. Avoiding
mathematical expression of this, it is at least clear that the ortho/para ratio
comes to zero at 0 °K. This corresponds to pure parahydrogen, the form
that can get rid of its rotational energy in the $J = 0$ state—orthohydrogen
cannot because the lowest state open to it is $J = 1$.

Calculations on this basis still failed to reproduce the experimental $C_V(T)$
plot for hydrogen, but the discrepancy vanished, and good agreement was
secured on the assumption that equilibrium is *not* established between the
nuclear spin isomers. Molecular hydrogen, prepared at room temperature,
retains its original 3 to 1 ortho/para ratio when progressively cooled. The
interconversion has a low probability (half-life about 3 years at S.T.P.),
but is susceptible to catalysis—particularly by active charcoal at low, but
not high, temperatures. This opened the way to the preparation of nearly
pure parahydrogen—by desorption of hydrogen from this catalyst at 20 °K.
Brought to normal temperatures, it remains unchanged almost indefinitely.
It differs significantly in properties from normal hydrogen (higher vapour
pressure, lower triple point), especially in thermal conductivity; a fortunate
circumstance exploited in determining ortho/para ratios.

Homogeneous conversion occurs at 600–1000 °C by a dissociative mechanism, providing a standard example of a reaction following $1\frac{1}{2}$ order
kinetics. Conversion is stimulated at normal temperatures by atomic hydrogen or hydrogenation catalysts—hardly surprisingly. This gives an indication
of the fields where the conversion is diagnostically useful. Paramagnetic
gases (O_2, NO, NO_2) and paramagnetic ions in solution (e.g. $Cu^{2+} \rightarrow Mn^{2+}$)
catalyse the conversion differently; lacking discussion, the guess can be
made that this is to do with otherwise forbidden ortho $\rightleftharpoons$ para transitions
under the influence of magnetic perturbation.

Nuclear spin isomerism, not unique to hydrogen, involves principles
of such wide implication that it is very desirable to follow the subject a little
closer to basic 'laws of Nature'. We have to accept the following as such:

1. Fundamental particles, and atoms or molecules built from them
are, kind for kind, indistinguishable from each other.

2. The behaviour of fundamental particles follows some wave equation.

3. Fundamental material particles—electrons, protons, neutrons—each
have angular spin momentum defined by a quantum number of $\frac{1}{2}$.

Conducting discussion in terms of two non-interacting electrons, let $\psi_a(1)$ and $\psi_b(2)$ be wave functions for electrons (1) and (2) in respective states a and b, each wave function being referred to its own set of space co-ordinates. A satisfactory solution to the wave equation for the joint system is the product $\psi_a(1)\psi_b(2)$. If the electrons are interchanged, the corresponding solution is $\psi_a(2)\psi_b(1)$. These two wave functions would be distinct from each other for the interchange of distinguishable particles. But electrons are not distinguishable, so that the two 'states' are not only of the same energy, but do not differ in any way. There is indeed only *one* state, and therefore these two wave functions, satisfactory in all other respects, offend against *quantum statistics*, because they imply distinguishability where there is none.

The difficulty is obviated by use of the principle that any linear combination of satisfactory solutions to the wave equation is itself a satisfactory solution. Then the linear combinations

$$\psi_S = \psi_a(1)\psi_b(2) + \psi_a(2)\psi_b(1) \tag{4.42}$$

$$\psi_A = \psi_a(1)\psi_b(2) - \psi_a(2)\psi_b(1) \tag{4.43}$$

for the joint system answer this description. Interchange of (1) and (2) does not change the sign of ψ_S, therefore called the symmetrical combination, but reverses the sign ψ_A, the antisymmetrical combination. This is not significant to ψ^2, the *probability function*. The indistinguishability principle can be tersely expressed by $\psi^2(1,2) \equiv \psi^2(2,1)$, and this identity is satisfied by $\psi(1,2) = \psi(2,1)$ or by $\psi(1,2) = -\psi(2,1)$. No distinction between the electrons is made by these equations, so that either is satisfactory in this respect—one or the other, but not *both*. This is because linear combination of the two leads back to the wave functions already rejected on the grounds that they assert distinguishability.

If, now, the *states* are made identical (e.g. by writing a for b in the above equations), it is seen that ψ_S remains finite, but ψ_A comes to zero, and so of course does the probability function, ψ_A^2. The antisymmetrical wave function therefore implies that no two electrons can occupy identical states. This is the Pauli exclusion principle, on which the periodic table depends— strong evidence, not lacking other support, that the *total* wave function describing the behaviour of electrons must be antisymmetrical. Other evidence shows the same to be true for protons and neutrons. All three fundamental material particles follow Fermi–Dirac statistics, which allow only single occupation of any state.

Like atomic nuclei are indistinguishable. They are composed of protons and neutrons, each with spin $\frac{1}{2}$, each obeying the Pauli principle. The exchange between states of like nuclei of mass number A therefore involves A interchanges between pairs of particles. Since each such interchange of fundamental particles (protons, neutrons) causes the wave function to change sign, the exchange of nuclei will make A changes of sign. If A is even, the wave function for the two nuclei will be symmetrical; if A is odd, it will be antisymmetrical. There is an associated rule about nuclear spin. The spin of a nucleus is a vector resultant of the spins of the protons and neutrons it contains. If A is even, the total nuclear spin must be integral, $i = 0, 1, 2, \ldots$; if A is odd, it must be half-integral, $i = \frac{1}{2}, \frac{3}{2}, \ldots$—what value in each case to be found by experiment.

Summarising, the total wave function for two nuclei of even mass number and zero or integral spin is symmetrical. Such nuclei follow Bose–Einstein statistics, which allow multiple occupation of any state by indistinguishable particles. On the other hand, the total wave function for two nuclei of odd mass number and half-integral spin is antisymmetrical. Such nuclei follow Fermi–Dirac statistics. A summarising statement is provided in Appendix 1.

These rules can now be applied to homonuclear diatomic molecules. Each nucleus, of spin i, has $(2i + 1)$ degenerate spin states (i.e. quantised projections of angular spin momentum on an external axis of reference). How are these $(2i + 1)$ states for each nucleus to be combined?

For simplicity, let $(2i + 1) = n$. There are n pairs of like spin states defined by n symmetrical wave functions. Pairs of unlike spin states must be formed by linear combinations of wave functions, symmetrical and antisymmetrical in equal number, namely,

$$\frac{n!}{2!(n - 2)!}$$

of each. For H_2, the wave functions are

		m_{sz}
$\psi_{+\frac{1}{2}}(1)\psi_{+\frac{1}{2}}(2)$ *		$+1$
$\psi_{-\frac{1}{2}}(1)\psi_{-\frac{1}{2}}(2)$ *	Symmetrical	-1
$\psi_{+\frac{1}{2}}(1)\psi_{-\frac{1}{2}}(2) + \psi_{+\frac{1}{2}}(1)\psi_{-\frac{1}{2}}(2)$ †		0
$\psi_{+\frac{1}{2}}(1)\psi_{-\frac{1}{2}}(2) - \psi_{+\frac{1}{2}}(1)\psi_{-\frac{1}{2}}(2)$	Antisymmetrical	0

* Not infringing the Pauli principal because rotation is yet to be considered. Vibration does not enter because it is not a function of orientation and ψ_{vib} is always symmetrical relative to interchange of nuclei.

† It is not apparent at a glance how this corresponds to a resultant nuclear spin of 1, but this is the case; on the other hand, the projections of the angular spin momentum, m_{sz}, on an external (z) axis of reference are shown against each wave function.

In general

$$\frac{n!}{2!(n-2)!} = \frac{n(n-1)}{2} = \frac{(2i+1)2i}{2} = i(2i+1)$$

The total number of symmetrical functions $= (2i+1) + i(2i+1)$

$$= (i+1)(2i+1)$$

and of the antisymmetrical functions $= i(2i+1)$

This gives the ratio symmetrical/antisymmetrical $= \dfrac{i+1}{i}$

For hydrogen, with $i = \frac{1}{2}$, this ratio is 3. For deuterium, with $i = 1$, it is 2. But deuterium, of mass number 2, follows Bose–Einstein statistics, and the total wave function for D_2 is symmetrical; symmetrical spin states therefore combine with 'J even' rotational states, and it is the ortho-form that is stable at 0 °K. For oxygen, $A = 16, i = 0$; the total wave function must be symmetrical and 'J odd' rotational states are forbidden—alternate rotational lines are missing from the O_2 spectrum, but appear in that of $^{16}O^{18}O$, with distinguishable nuclei. In the case of nitrogen, $A = 14, i = 1$, so the behaviour is similar to that of deuterium. The very much closer spacing of the rotational levels, however, means that except spectrally, the nuclear spin isomerism is not observable. This is so for all molecules as heavy or heavier—diatomic, or others to which similar symmetry considerations apply.

Attention may now be given to the standard entropy of normal hydrogen. Only pure parahydrogen can have zero residual entropy at 0 °K. What will be the residual entropy of hydrogen with 3 to 1, frozen-in ortho/para ratio?

Rotational states of degeneracy $(2J + 1)$ have to be considered to arrive at statistical weights. These will be:

para

$$i(2i+1)(2J+1) = 1 \text{ for } i = \tfrac{1}{2} \text{ and } J = 0$$

ortho

$$(i+1)(2i+1)(2J+1) = 9 \text{ for } i = \tfrac{1}{2} \text{ and } J = 1$$

In effect, we have to consider a mixture of 10 kinds of hydrogen in the solid state at 0 °K, $\frac{1}{4}$ para- and $\frac{3}{4}$ ortho- divided equally between 9 states. The residual entropy due to this mixed-upness can be calculated using $S_0^\circ = k \ln W$, with

$$W = \frac{N!}{\dfrac{N}{4}!\left(\dfrac{N}{12}!\right)^9}$$

where N is Avogadro's number. Using Stirling's approximation that $\ln N! = N \ln N - N$, this leads to

$$S_0^\circ = R(\tfrac{1}{4} \ln 4 + \tfrac{3}{4} \ln 12) = 4 \cdot 39 \text{ cal } {}^\circ\text{K}^{-1} \text{ mole}^{-1}$$

As will be shown, this is an 'entropy of mixing'.

It might then be expected that this S_0°, added to the normally determined, calorimetric entropy, would give an acceptable S_{298}°. In 1930, this sum was $29 \cdot 69 + 4 \cdot 39 = 34 \cdot 08 \text{ cal } {}^\circ\text{K}^{-1} \text{ mole}^{-1}$, in reasonable agreement with $33 \cdot 98 \text{ cal } {}^\circ\text{K}^{-1} \text{ mole}^{-1}$ found from statistical mechanical calculation. Later work revealed a very outstanding C_P peak below the lowest temperature of the earlier measurements—for normal, but not for parahydrogen. Molecular rotation multiplicity due to alignment of molecules is apparently absent in solid hydrogen at $0\,{}^\circ\text{K}$, but comes in with rising temperature at about $2\,{}^\circ\text{K}$. This should not seriously affect the calorimetric S_{298}°, because it transfers an entropy increment from S_0° to $\int C_P/T.dT$.

Accepting the statistical mechanical value of $33 \cdot 98 \text{ cal } {}^\circ\text{K}^{-1} \text{ mole}^{-1}$ as the better, it is still not suitable for combining with the S_{298}° values for other substances. All the tabulated S_{298}° data are *practical entropy values* from which the effects of multiplicity of nuclear spin orientation (as well as of isotope mixing) are excluded. They must therefore be excluded in the case of hydrogen, and this means rejecting from the S_0° calculated above the terms $R(\tfrac{1}{4} \ln 4 + \tfrac{3}{4} \ln \tfrac{4}{3}) = 1 \cdot 118 \text{ cal } {}^\circ\text{K}^{-1} \text{ mole}^{-1}$ for ortho-para mixing, and $\tfrac{3}{4} R \ln 3 = 1 \cdot 637 \text{ cal } {}^\circ\text{K}^{-1} \text{ mole}^{-1}$ for nuclear spin multiplicity of the ortho-molecules. The necessary deduction, $2 \cdot 755 \text{ cal } {}^\circ\text{K}^{-1} \text{ mole}^{-1}$ can also be arrived at from the $(2i + 1)$ multiplicity of spin states per proton in H_2; this leads to $2R \ln 2 = 2 \cdot 755 \text{ cal } {}^\circ\text{K}^{-1} \text{ mole}^{-1}$.

Hence the practical S_{298}° of normal hydrogen is $33 \cdot 98 - 2 \cdot 76 = 31 \cdot 22$ cal ${}^\circ\text{K}^{-1}$ mole^{-1} (the most recent figure is $31 \cdot 21$ cal ${}^\circ\text{K}^{-1}$ mole^{-1}). The validity of this result is confirmed by numerous cross-checks of the form $\Delta S^\circ = \Sigma n S^\circ$, for reactions involving hydrogen; ΔS° is determinable independently of any S° data, so a value for the standard entropy of hydrogen can be found when it is the only unknown in the algebraic summation represented by $\Sigma n S^\circ$.

A final comment on practical (sometimes called virtual) entropies is needed. Fundamentally, it could be argued, nuclear spin entropies should be included in the summations giving rise to the standard data. But in the practical politics of chemical reactions (barring the exceptional cases discussed), this would not be significant because all the nuclear spin entropies would cancel; there is no change in the number or nature of the nuclei in a normal chemical reaction. It nevertheless remains essential to take account of the basic symmetry rules in the statistical mechanical process of counting molecular rotational states, i.e. in evaluating the rotational partition function. Thus, a molecule X—Y, rotated 180° about an axis perpendicular to its length, is brought to a state, Y—X, different from the first. But for a homonuclear molecule X—X this is not so; the rotation has interchanged indistinguishable nuclei, and there is no new state. In general, the sum of states must be divided by a symmetry number, σ; 1 for X—Y, 2 for X—X. The same consideration enters for all polyatomic molecules; σ is 2 for H_2O, 3 for NH_3, 12 for CH_4—if there are n equivalent orientations of a molecule in space, only $1/n$ of them may be counted.

4.8   A conspectus of some standard entropies

To gain some appreciation of the magnitudes of the entropies of substances and to see some of the factors contributing to them, it is useful to survey a selection of S°_{298} data such as displayed in Table 4.3.

For gaseous substances, it is the disorder due to random translational motion that provides by far the largest contribution—the translational entropy, S_t°—to the total standard molar entropy, S°. For the noble gases, and all other monatomic gases or vapours from halogens to metals, it is the *only* contribution, and is seen to increase regularly with increasing mass. Indeed, from one monatomic gas to another at the same temperature and pressure, mass is the only significant, entropy-determining variable. Statistical mechanics provides the Sackur–Tetrode equation.

$$S_t = \tfrac{3}{2}R \ln M + \tfrac{5}{2}R \ln T - R \ln P - 2 \cdot 314$$

or

$$S^\circ_{t,298 \cdot 15} = \tfrac{3}{2}R \ln M + 25 \cdot 992 \text{ cal } {}^\circ K^{-1} \text{ mole}^{-1}$$

(4.44)

where M is molecular weight and the numerical terms are functions of fundamental constants. It is the solutions to this equation that are in splendid concordance with appropriate 'calorimetric' entropies. Gratuitous though it may be, the remark must be made that this equation expresses the quantisation of translational energy, which is, of course, a most fundamental fact. Were this energy continuously variable, it would offer an infinity of states, and no other kind of state would stand a significant relative chance of being occupied.

For gases with diatomic molecules, S_t°, already greater by reason of mass, is supplemented by rotational entropy, S_r°. This, as already discussed, is larger for hetero- than for homonuclear molecules (see the isotopic variants of H_2; Br_2, IBr, I_2—also the triatomic light and heavy waters). Vibrational entropy, S_v° is zero for light diatomic molecules, entering only at higher temperatures when kT becomes commensurate with the larger vibrational

Table 4.3 Some standard entropy data, $S^\circ_{298.15}$, cal $^\circ K^{-1}$ mole^{-1}

He	Ne	Ar	Kr	Xe	Rn
30·13	34·95	36·98	39·19	40·53	42·10

Li(g)	Na(g)	K(g)	Rb(g)	Cs(g)	Cu(g)	Ag(g)	Au(g)
33·14	36·71	38·30	40·63	41·94	39·74	41·32	43·12
Li(c)	Na(c)	K(c)	Rb(c)	Cs(c)	Cu(c)	Ag(c)	Au(c)
6·70	12·24	15·34	18·1	20·1	7·97	10·20	11·31
(26·4)	(24·5)	(23·0)	(22·5)	(21·8)	(31·8)	(31·1)	(32·8)

H	H_2	D	D_2	T	T_2	HD	DT
27·39	31·21	29·46	34·62	30·66	36·63	34·34	37·04
	S_t 28·081		S_t 30·144			S_t 29·288	
	S_r 3·047		S_r 4·424			S_r 4·994	
	S_v 0·000		S_v 0·000			S_v 0·000	

$H_2O(g)$	$D_2O(g)$	$HDO(g)$	$T_2O(g)$	$DTO(g)$		
45·11	47·38	47·66	48·79	49·46	$NH_3(g)$	46·02
$H_2O(l)$	$D_2O(l)$				$PH_3(g)$	50·29
16·73	18·08				$AsH_3(g)$	53·22
(28·4)	(29·3)				$SbH_3(g)$	55·61

				HF	HCl	HBr	HI
				41·47	44·64	47·48	49·63
F(g)	Cl(g)	Br(g)	I(g)				
37·92	39·46	41·81	43·18				
$F_2(g)$	$Cl_2(g)$	$Br_2(g)$	$I_2(g)$	IBr(g)			
48·49	53·20	58·65	62·28	61·8			
S_t 36·834	38·649	41·117	42·495	41·89			
S_r 11·507	14·042	16·207	17·744	18·30			
S_v 0·114	0·528	1·306	2·004	1·62			
		$Br_2(l)$	$I_2(c)$				
		36·4	27·76				
		(22·3)	(34·5)				

N_2	45·77
O_2	49·01
CO	47·30
NO	50·34
CO_2	51·08
NO_2	57·36
N_2O	52·55
SO_2	59·29

Table **4.3** (*contd.*)

	CCl₄(g) 73·8			Li(c) 6·70 LiH(c) 5·9		

$$
\begin{array}{lllllll}
 & \text{CCl}_4(g)\;73{\cdot}8 & & & \text{Li(c)}\;6{\cdot}70 & \text{LiH(c)}\;5{\cdot}9 & \\
 & \text{SiCl}_4(g)\;79{\cdot}1 & & & \text{Be(c)}\;2{\cdot}28 & \text{BeO(c)}\;3{\cdot}37 & \\
\text{TiCl}_4(g) & 84{\cdot}3 & \text{GeCl}_4(g)\;83{\cdot}0 & \text{B(c)}\;1{\cdot}40 & \text{B}_2\text{O}_3(c)\;12{\cdot}91 & \text{B}_2\text{O}_3(gl)\;18{\cdot}6 & \\
\text{ZrCl}_4(g) & 86{\cdot}5 & \text{SnCl}_4(g)\;87{\cdot}2 & \text{C(diamond)}\;0{\cdot}585 & & & \\
\text{HfCl}_4(g) & 89{\cdot}8 & \text{PbCl}_4(g)\;89{\cdot}4 & \text{C(graphite)}\;1{\cdot}36 & & & \\
\end{array}
$$

Na(c) 12·24

Mg(c) 7·81 MgO(c) 6·55 Mg(OH)$_2$(c) 15·09 MgCO$_3$(c) 15·7

MgSO$_4$ 21·9 MgSO$_4$.2H$_2$O 30·2 MgSO$_4$.6H$_2$O 83·2

MgCl$_2$ 21·4 ... H$_2$O 32·8 ... 2H$_2$O 43·0 ... 4H$_2$O 63·1 ... 6H$_2$O 87·5

Al(c) 6·77 Al$_2$O$_3$(α) 12·18 Al$_2$(SO$_4$)$_3$ 57·2 KAl(SO$_4$)$_2$ 48·9

$\qquad\qquad\qquad\qquad\qquad\qquad$ KAl(SO$_4$)$_2$.12H$_2$O 164·3

Si(c) 4·51 SiC(hex) 3·94 SiO$_2$(α quartz) 10·00 SiO$_2$(gl) 11·2 SiO(g) 50·5

P (white) 9·80 P (violet) 7·4 P (black) 5·3

S(α) 7·62 S(β) 7·78

Cu(c) 7·97 Cu$_2$O(c) 22·4 CuO(c) 10·19 CuSO$_4$ 25·3 CuSO$_4$.H$_2$O 33·0

$\qquad\qquad\qquad\qquad\qquad\qquad$ CuSO$_2$.3H$_2$O 52·4 CuSO$_4$.5H$_2$O 70·2

Zn(c) 9·95 Zn(g) 38·45 (28·5) ZnO(c) 10·43 ZnS(c) 13·8

$\qquad\qquad\qquad$ ZnSO$_4$ 27·0 ZnSO$_4$.6H$_2$O 86·9 ZnSO$_4$.7H$_2$O 92·9

Hg(l) 18·71 Hg(g) 41·80 (23·1) H$_2$O (red) 16·80 HgO (yellow) 17·3

Ti(c) 7·30 TiC(c) 5·79 TiN(c) 7·24 TiO$_2$ (rutile) 12·04

Sn (white) 12·20 Sn (grey) 10·55

Pb(c) 15·49 PbO (red) 15·6 PbO$_2$(c) 18·3 Pb$_2$O$_3$(c) 36·3 Pb$_3$O$_4$(c) 50·5

W(c) 7·14

Fe(α) 6·49 Fe$_{0.947}$O(c) 13·74 Fe$_2$O$_3$(c) 20·90 Fe$_3$O$_4$(c) 35·0

$\qquad\qquad\qquad\qquad\qquad$ Fe(SO$_4$)$_2$.7H$_2$O 97·8 FeCl$_3$(c) 32·2 FeCl$_3$(g) 86·9

Ni(c) 7·14 NiO(c) 9·08 Ni(CO)$_4$(g) 97·1 Ni(CO)$_4$(l) 74·9 (22·2)

Notes: (g)=(ideal) gas; (l)=liquid; (c)=crystalline (sometimes omitted when obvious); (gl)=glassy. Probable errors are omitted for clarity; they are not significant to the discussion in the text.

quanta. Its contribution to the gaseous halogens is seen to be significant, increasing with mass. Increasing molecular complexity increases molar entropy, not only by the mass effect on all three entropy contributions, but also by the incidence of additional, internal modes of motion. Within a group of analogues, such as the group IV tetrachlorides, the entropies are remarkably regular in trend.

Less regularity is to be expected in condensed phases, where the large and regular contribution of S_t° (depending, paradoxically, on randomness) is mainly (liquids) or wholly (solids) lost. In the table some entropy differences between gaseous and condensed phases are shown in brackets. For the evaporation of liquids, still characterised by molecular disorder, ΔS° is fairly uniform in value (Br_2, 22·25; CCl_4, 22·5; $Ni(CO)_4$, 22·2; Hg, 23·1 cal $^\circ K^{-1}$ mole^{-1}), except where there are good grounds for believing that the liquid state entropy is depressed by some peculiar element of ordered structure (H_2O, 28·4; D_2O, 29·3 cal $^\circ K^{-1}$ mole^{-1}). Trouton's classical rule will be recalled, and deviations from it shown by 'associated liquids'.

The entropies of solids are very much more specific to the structure, and the kind and strength of the forces maintaining crystalline order. The very small entropy of diamond, a three-dimensional, covalent giant molecule, is outstanding. Graphite, gigantic in but two dimensions, has somewhat greater entropy. Trends arising from lightness and hardness can be seen in the earlier members of the two short periods. The dominance of strong structural forces over entropy appears repeatedly in very hard or refractory substances, elements (W, cf. Pb) or compounds: a compound may for this reason be of lower entropy than its 'parent' element (Li, LiH; Mg, MgO; Si, SiC; Ti, TiC, TiN). Because the entropies of solids are so dependent on structure, there is no rule, parallel to Trouton's, applying to the generally smaller and very variable latent heats of melting ($\lambda_M = T_M \Delta S_M$).

Emphasis on the high entropy of gases must not be pressed too far. It would, of course, be absurd to suppose that the entropy of any gas is greater than that of any liquid or solid under comparable conditions (although it can be made so by sufficient reduction of pressure). It is to be remembered that when a freely moving particle is captured by a field of force, the loss in S_t is offset by gain of S_v, since the particle must vibrate in its new, constrained situation. Since the entropy of transfer from condensed to gaseous phase (at 1 atm) is, by and large, roughly constant, the difference between the molar entropies in the two states becomes proportionately smaller with increasing molecular mass and complexity (H_2O, 170%; $Ni(CO)_4$, 30%). Even neither very heavy nor very complex substances can have large solid state entropies—considerably augmented by water of crystallisation. It is of interest that the entropy contribution per molecule of water of crystallisation, although variable, averages to about 9·4 cal °K^{-1} mole^{-1}—comparable with that of ice (9·91 cal °K^{-1} mole^{-1} at 0 °C).

Two other features illustrated in the table are that the form of a polymorphic substance stable at the higher temperature has the higher entropy (S, Sn), and that the entropy of the glassy state is greater than that of the crystalline state (B$_2$O$_3$, SiO$_2$)—exemplifying points previously made.

Questions may now be put to the reader. Has this kind of review of the data carried a message that entropies of substances are properties of chemical interest? Should it not be part of chemical know-how to regard a given reaction, and, without consulting tables, to make an informed guess about its $\Delta S°$, thus arriving at some idea of the entropic contribution (plus or minus) to its affinity, *and* of the likely temperature coefficient of its affinity?

It was said in introductory discussion that thermodynamic data sometimes shout that here is something interesting to be investigated by all appropriate means. The occasion to illustrate this, even if prematurely, cannot be resisted. Attention is therefore directed to Table 4.4, containing some conventional ionic standard entropies, such as are tabulated and used along with the other $S°_{298}$ data. Unlike the other data, they make no pretence of being absolute, but are all relative to an arbitrary zero assigned to a mole of hydrogen (hydronium) ions in hypothetically ideal, unimolal, aqueous solution. Then the figures may be interpreted as the entropy change that would result from the replacement of a mole of hydrogen ions in such a solution by a mole of the other kind of ion in question.

Table 4.4   Some conventional standard entropies, $S°_{298.15}$ of ions in ideal unimolal aqueous solution, cal °K^{-1} mole^{-1} (ref. 10)

				H$^+$	0.0		
Li$^+$	4·7	Mg^{2+}	−32·7	Al^{3+}	−76·0	F$^-$	−2·3
Na$^+$	14·0	Ca^{2+}	−11·4	Fe^{3+}	−61·0	Cl$^-$	13·5
K$^+$	24·2	Sr^{2+}	−7·3	Pu^{4+}	−87·0	Br$^-$	19·7
Rb$^+$	28·7	Ba^{2+}	2·3	NO$_3^-$	35·0	I$^-$	25·3
Cs$^+$	32·0	Fe^{2+}	−25·9	SO$_4^{2-}$	4·4	OH$^-$	−2·5

Very large entropy effects of both signs are seen, varying in a systematic way with ionic charge and radius. Most of the ions shown are monatomic and, in the ideal gas phase, would have uninteresting entropies calculable by the Sackur–Tetrode equation. How can transfer to the aqueous medium produce such dramatically differentiated changes? Clearly, no *intraionic* effects can be responsible, so the reasons must be sought in what the ions are doing to the water. It is understandable that the extremely intense electrostatic field (to be measured in 10^7 volt cm^{-1}) of a small cation of multiple charge should 'freeze' polar water molecules into an immobile hydration shell—with consequent loss of freedom of motion and entropy. But it is notable that the entropy loss can be very great. On the other hand, why does an ion like Cs^+ apparently cause a substantial entropy rise? Of course, the figures only have relative significance, and this might be illusory. If, however, a salt containing two such entropy-enhancing ions (e.g. CsI) is examined, it is found that when it is dissolved in water, it decreases the viscosity. This is clear evidence that these ions do indeed in some way loosen up the structure of the water, which is of course unique among solvents for its degree of 'structuredness'. Sufficient at this stage to remark that entropy, along with other properties, plays an essential part in the study of ionic and non-ionic solute-solvent interactions—a field of great current interest.

It would be inappropriate to omit mention of the organic chemical field, where the application of thermodynamic functions is wide and profitable. For entropies, with the other extensive properties, the best use is made of all sources of information, including spectroscopic data used in statistical mechanical calculations. Since interpretation of such data becomes difficult, and the calculations complex, for large organic molecules, recourse is had to semi-empirical correlation methods of every kind that ingenuity can devise. Names of the methods—that of structural similarity, or of group contributions, give an inkling of the direction of such efforts. The source of

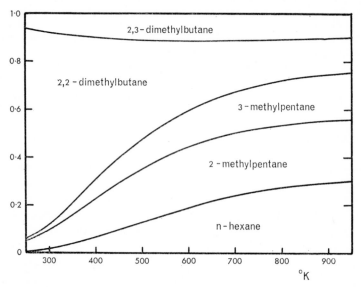

Fig. 4.4 Equilibrium mole fractions of hexane

their success (within appropriately widened limits of error) can be seen in the repetitive occurrence of common structural features, and of other common factors (symmetry numbers, restricted rotations, steric repulsions) in the overwhelmingly great number of organic compounds. There is an excellent monograph on this subject,[11] which cannot now be followed, except to make a point of quite general importance, as follows.

All the data presented in this section relate to 25 °C. They are, of course, adapted for application at other temperatures by methods, some already outlined, others yet to be discussed, involving the ubiquitous heat capacities (measured, fundamentally calculated or estimated by correlations). Without the necessary data for proper calculation, the semi-quantitative kind of mental exercise the writer has advocated must be cautious. Fig. 4.4 presents an interesting example of how a situation may be radically altered by change of temperature; it represents the *equilibrium* proportions of the hexanes in dependence on temperature.[12] Without detailed discussion, it clearly bears thinking about.

REFERENCES

1. Lewis, G. N., and Randall, M., *Thermodynamics*, McGraw-Hill, New York, 1923, p. 448.
2. See Ref. 1 above, p. 445.
3. Craig, R. S., Krier, C. A., Coffer, L. W., Bates, E. A., and Wallace, W. E., *J. Amer. Chem. Soc.*, 1954, **76**, 238.
4. Pauling, L., *J. Amer. Chem. Soc.*, 1935, **57**, 2680.
5. Bjerrum, N., *Science*, 1952, **115**, 385.
6. Moore, W. J., *Physical Chemistry*, 4th ed., Longmans, London, 1963, chap. 16.
7. Ubbelohde, A. R., *Melting and Crystal Structure*, Clarendon, Oxford, 1965.
8. Bernal, J. D., *Proc. Roy. Soc.*, 1964, **A, 280**, 299; *Scient. American*, no. 267, August 1960.
9. Eisenberg, D., and Kauzmann, W., *The Structure and Properties of Water*, Clarendon Press, Oxford, 1969; Ives, D. J. G., and Lemon, T. H., *Roy. Inst. Chem. Rev.*, 1968, **1**, 62.
10. Kelley, K. K., and King, E. G., Bulletin 592, Bureau of Mines, Washington, 1961.
11. Janz, G. J., *Thermodynamic Properties of Organic Compounds*, Academic Press, New York, 1967.
12. Rossini, F. D., in *The Chemical Background to Engine Research*, ed. Burk and Grummit, Interscience, New York, 1943, p. 55.

5

Some Basic Arguments and Relationships

5.1   Introduction

In this chapter, a re-examination of the concept of equilibrium and the disposal of the difficulties surrounding the second law precede the development of some working equations. These equations are applied to equilibria in closed systems. The recording of data is considered.

5.2   Equilibrium

It has already been emphasised that 'equilibrium' has only relative significance. A coin standing on edge or lying flat is in equilibrium states of lesser or greater stability. In the first, it recovers only from the gentlest touch: in the second, it falls flat again after any rough usage. A general description of an equilibrium state, covering all degrees of stability from fragile to massive must therefore be made in terms of response to an *infinitesimal* disturbance. 'A state stable relative to all infinitesimally differing states' is a start. It covers smaller or greater maxima of stability, considered as a function of 'state', but it needs developing into more specific forms to provide generally valid tests for, or criteria of, equilibrium. For the trivial example of the coin, this development might be potential energy, E, expressed as a function of angle to the horizontal, θ (Fig. 5.1a). Minima in potential energy ($dE/d\theta = 0$), coincide with maxima in stability, and respond to the test that $dE = 0$ for an infinitesimal disturbance $d\theta$. For a finite but small disturbance, $\Delta\theta$, $\Delta E > 0$, but, on abatement, E spontaneously rolls back to its minimum. For too large a disturbance, E may ride over a maximum, so that the system may be kicked downstairs into quite a different, lower state. This is why, in the limit, it is necessary to stick to infinitesimals in a general definition.

Second thoughts (about a cascade of coins on to the floor) might suggest that a coin standing on edge is seldom found. Perhaps consideration of a

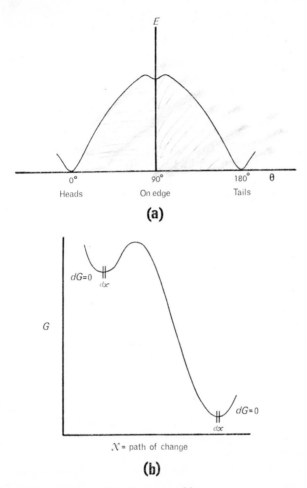

Fig. 5.1 Potential energies of a coin on a table

large number of randomly assembled coins should bring in some aspect of probability.

Without more ado, it may be granted that Gibbs free energy, G, is pre-eminently suitable to quantify stability *in macroscopic systems*. For a system changing spontaneously, G is decreasing. When the system reaches equilibrium and tends to change no further, G is at a minimum. Then, for any infinitesimal displacement along the path of change, $dG = 0$, as indicated in Fig. 5.1b—so drawn to make a significant point. In chemical thermodynamics we have to consider *states* of a system differing chemically and often very widely in all extensive properties; the physicist would call them different systems. This leads to difficulty of communication between the disciplines, and is part of the reason why there is a bogey to be laid before presenting formal derivations of precise criteria of equilibrium.

5.3   The second law

An *ad hoc* survey of a dozen texts turned up twelve versions of this law—variations on themes by Clausius, Thomson (Lord Kelvin), Planck and Maxwell. The impression might be forgiven that all of them skirt around some theorem too esoteric to be framed in a few easily understood words. Pointing in the same direction is the fact that more than a century's worth of voluminous writing on the subject has spilled over into mathematics, philosophy, even demonology.* It is hard lines on hard-pressed students, shovelling about to find the nub, to encounter 'It is, perhaps, the most extraordinary natural law with which science is yet acquainted, and its understanding demands much thought', and 'Most students of chemistry find difficulty in appreciating its meaning and content', and, to cap all, '. . . in the last analysis, the most interesting thing about the second law of thermodynamics is that it is not absolutely true'. One can feel sympathy with the student who says 'To hell with that', and pushes off to the pub. The job now is to find the nub and see whether it can be expressed in few words.

To begin, one of the Clausius statements, 'Heat cannot of itself pass from a colder to a hotter body' is easily accepted—provided we know the difference between hot and cold. Pressed as to why we accept this, we reply that such a flow of heat is not a natural process and never occurs spontaneously. Pressed further, we fall back on saying that in our experience and belief the statement is true; we have to start our thinking somewhere, and this seems to be as basic a starting-point as any. We also accept the inverse statement 'Heat can of itself pass from a hotter to a colder body'. This *is* a natural process, and we further believe that, like all such processes and *only* such, it can, in principle, be used to provide work.

Another statement[2] is 'No process has ever been discovered whose sole result is the transfer of energy from thermal to non-thermal forms'. We perceive, in terms of the previous statement, that if heat passing from the hotter body were all siphoned away into work, none would be left to pass

* 'Maxwell's demon', by manipulating a shutter in a partition between two compartments, can separate hot gas molecules from cold ones—or can he not? The writer, far from decrying the intellectual exercises of his betters, draws attention to a recent article on Maxwell's familiar.[1]

on to the colder body, so there would no longer be a natural process. We are therefore able to believe this statement as well—even if we had not already traced out Carnot's argument. Are we therefore getting round to the view that the essence of the second law is that to provide useful work a process must be spontaneous? Surely, there must be something more in it than this?

Taking the bull by the horns, we look at another Clausius statement: 'In an isolated system the entropy tends to a maximum value.' This will satisfy those with an aversion to the obvious. The student (one eye on the pub) might sourly comment that it is the kind of over-concise statement that is unnecessary to those who understand the subject, and unintelligible to those who do not. Agreed. The next step is therefore to examine it word by word, and judiciously expand it.

An isolated system has no communication with its surroundings; in effect, it has none. It can neither gain nor lose any kind of energy. The energy content of an isolated system is constant. The over-concise statement therefore excludes from consideration one factor we have hitherto regarded vital in relation to the approach to equilibrium, namely, the minimisation of energy. Why does it do this? To focus attention on entropy.

The over-concise statement uses the word 'tends'. There is something conditional about this, even slightly shifty, so that, wishing to ask questions, we do not know how to frame them. Suppose we expand the statement into the form 'If an isolated system is not in equilibrium, a change towards equilibrium will tend to occur, and any such change will be accompanied by an increase in entropy.' This does not transgress the terms of the shorter statement, but allows a question to be asked. Is the tendency to change always to be attributed solely to the intrinsic property of entropy to increase to a maximum value, or, depending on the nature of the initial disequilibrium, has it sometimes some other primary cause?

At this stage, being stupid chemists, we think it best to consult a physicist. We put the question to him, asking as well what kinds of initial disequilibria has he in mind for *his* isolated systems? He will probably reply that he thinks, for example, of initial inequalities of pressure or temperature within the system, but adds that we have no right to ask questions about cause and effect, because they are extrathermodynamic, i.e., outside the province of the subject. In any case, since energy cannot come into it, the only possible thermodynamic answer is that it is strictly a matter of maximisation of entropy. The statistical mechanical view confirms this; $S = k \ln W$, the system moves towards a state of maximum probability, and that is all.

Impressed, we next ask, if this is really the basic statement of the second law, which we are told is universally valid, how do we apply it to the non-isolated systems that interest us? The reply will be that we must combine our non-isolated system with its surroundings—all of them that can be significantly affected by anything happening in the system, if necessary taking in the whole of the universe. The new super-system can be considered as isolated, and we can apply the Clausius statement to *it* as it stands. If we really want to worry about causes, just let us keep $S = k \ln W$ in mind.

Returning to our own territory, we begin to think out the lesson in our own terms. Suppose our isolated system is a 2:1 mixture of hydrogen and oxygen. If its tendency to change is realised, the final state (with no energy lost) seems more fantastic than probable. We can, however, see that the entropy of extensively dissociated steam at several thousand degrees C is likely to be high in value. Given the necessary data, we could do the calculation, but need not bother. The exercise is unsuitable for us, and it will be better to use the second option allowed us by the physicist.

To do this, we drastically modify the Clausius over-concise statement to suit our purpose, as follows.

If a non-isolated system is in disequilibrium, a change towards equilibrium will tend to occur. If it does occur, it will be accompanied by an increase in entropy, provided that the entropy change in the system itself, *and* the entropy change in the surroundings are added together to obtain the significant total. In other words, if *all* the entropy changes, *wherever they are located*, consequent on a spontaneous change are added up, the total is always positive.

This revised statement can now be examined in terms of a chemical reaction allowed to proceed intelligibly—in the first instance, irreversibly (as in a calorimeter) at constant temperature and pressure. Care is taken to avoid the bugbear of confusion between system and surroundings by writing the often-used equation $\Delta H_{sys} = \Delta G_{sys} + T\Delta S_{sys}$ with subscripts to leave no doubt that all the 'delta terms' represent changes in extensive properties *of the system itself*. These changes are, of course, the same whether the reaction to which they relate is carried out irreversibly or reversibly—they are determined only by final and initial states.

Heat is transferred at constant pressure between the system and the surroundings. If the system loses enthalpy, the surroundings gain it, and vice versa. Then, $\Delta H_{surr} = -\Delta H_{sys}$. The entropy change in the surroundings, ΔS_{surr}, is equal to the heat they receive divided by the absolute temperature at which the heat is received,* i.e., $\Delta S_{surr} = \Delta H_{surr}/T$. We are told that,

---

* If, as we are considering, this heat is received at constant temperature, T, it is being received reversibly. But if it is not, because temperature difference is temporarily generated, this does not matter. The final ΔS_{surr}, when T is regained throughout, is dependent only on final and initial states.

if the reaction is spontaneous, $\Delta S_{sys} + \Delta S_{surr}$ must be positive. Therefore, $\Delta S_{sys} + \Delta H_{surr}/T$ must be positive. But $\Delta H_{surr} = -\Delta H_{sys}$, so $\Delta S_{sys} - \Delta H_{sys}/T$ must be positive, and so must $T\Delta S_{sys} - \Delta H_{sys}$ be. This is seen to be identical with $-\Delta G_{sys}$, so all this boils down to the statement that for a spontaneous reaction ΔG_{sys} must be negative. This is in line with our previous chemical-thermodynamic thinking.

Suppose, instead, the reaction were to be conducted reversibly. For this, it must be opposed by a 'work-absorbing load' that balances the tendency of the reaction to proceed, and establishes an equilibrium.* Adjustment of the load allows the reaction to proceed (or to be reversed) by infinitesimal increments with, in the limit, no disturbance of the equilibrium state. Then *non-thermal* energy is being transferred between the system and the surroundings. If the system loses Gibbs free energy, the surroundings gain it, and vice versa, i.e., $dG_{surr} = -dG_{sys}$. On the understanding that the equilibrium state is maintained, this equality can be multiplied by any factor, so that we can write $\Delta G_{surr} = -\Delta G_{sys}$. Since ΔG_{sys} is thus fully committed, the *heat* that can be transferred at constant pressure between system and surroundings is no longer equal to ΔH_{sys}; it is depleted by the measure of ΔG_{sys}, and only $T\Delta S_{sys}$ remains. If the system loses this amount of heat, the surroundings gain it, so it is evident without further argument that $\Delta S_{surr} = -\Delta S_{sys}$.

Looking now at the super-system (system + surroundings), we find $\Delta G_{total} = \Delta G_{sys} + \Delta G_{surr} = 0$ and $\Delta S_{total} = \Delta S_{sys} + \Delta S_{surr} = 0$, consistent with the maintenance of equilibrium in the super-system (G at a minimum, S at a maximum), which, regarded as isolated, toes the second law line. If all the non-thermal energy received by the surroundings (otherwise, optimum external work) were to be frittered away into heat at temperature T, we should get back to the irreversible reaction.

It seems, therefore, that there are two different ways of looking at the same thing, and the link between them is set out schematically in Fig. 5.2. Our chemical-thermodynamic way has been to attribute overriding significance to the *difference* between two quantities that *belong together* ($\Delta H_{sys} - T\Delta S_{sys}$), at the same time finding significance in separate consideration of them. The other way attributes significance only to the *sum* of two quantities that do not belong together ($\Delta S_{sys} + \Delta S_{surr}$).

Before proceeding further, let us consider whether we are indeed naughty, as alleged by the imaginary physicist, in asking questions about causes. We take an example, familiar to chemists, of a system initially in disequilibrium proceeding spontaneously towards equilibrium. It is a supersaturated solution of a salt that has been seeded and is crystallising. It is particularly

* An easy matter in electrochemical systems.

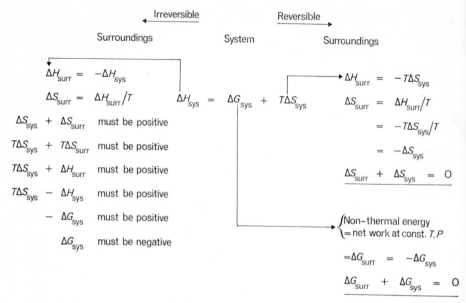

Fig. 5.2 The two thermodynamic viewpoints

striking to watch under the microscope; crystal edges, sharp and perfect in geometrical regularity, can be seen advancing into the solution phase. Why is this happening? It is because interionic forces are being satisfied, with fall of potential energy, and are building up a beautifully ordered space-lattice. This must involve a considerable entropy decrease, so that ΔS for the system in which the spontaneous process is proceeding must be negative. But the lost potential energy appears as heat, and is transferred to the surroundings, which thereby increase in entropy. Addition of the two entropy changes gives a positive result, so the second law is not infringed. Are we to say that all this happens simply because of the consequent over-all entropy rise, in conformity with what Clausius said? If we must keep our minds on $S = k \ln W$, we have to bring in the probability of the Universe, a concept so intangible as not even to qualify for the description of nonsense.

A chemical-thermodynamic stand must be taken.

It is legitimate to ask about causes and reasons. The reasons for the occurrence of a spontaneous change in a system are to be looked for within the system and not outside it. Only then is it possible to discriminate between energy and entropy effects, and see the balance between them in operation throughout chemistry and physics.

The necessity for such a declaration may seem remarkable. It can be traced to the subjects—engineering, physics, chemistry—that have, in historical sequence, provided the main context for the development of thermodynamics. There has been inevitable carry-over of tradition, to the detriment of the exposition, at large, of chemical thermodynamics. Different, if not inconsistent, languages are in use, sustained by differences in fields of interest as well as of tradition. Physics, concerned with physical changes and equilibria, need not worry about the relatively enormous energy changes accompanying chemical reactions—as great at 0 °K as at any other temperature. Chemistry has to, and it is hardly surprising that it should require its own formulation of the second law. The question is, what?

The equation $\Delta G = \Delta H - T\Delta S$ has been, but seldom is, referred to as the 'second law', but this needs some elucidation, as above. The statement 'Every spontaneous process generates entropy in one way or another' is not misleading, but neither is it satisfactory—it does not meet the chemists' needs and panders to tradition. Try as one may, every attempt to express the desired nub of the matter in a few words turns out to show but an aspect or a consequence. Let us therefore try an entirely new tack.

Although it is impracticable to discard the expression 'the laws of thermodynamics' because it is so firmly entrenched by tradition, it is better to regard these enunciations not as 'laws', but as *definitions*—of a kind to include the intrinsic properties of the quantities defined, such as the property of energy to be conserved. In this sense, the 'first law' defines energy; the 'second law' defines temperature and entropy; the 'third law' completes the definition of entropy.

The definitions comprising the second law, stemming from Carnot, Thomson and Clausius, are summarised in the following relationships, set out with minimal comment, but intelligible in the light of preceding discussion (e.g., section 4.5).*

For any reversible cycle,

$$\frac{w_{max}}{q_1} = \frac{T_1 - T_2}{T_1} \to 1 \text{ as } T_2 \to 0,$$

defining a zero of temperature independent of any particular substance or system.

$w_{max} = q_1 - q_2$ because energy is conserved.

* In terms of entropy—now to be used to define entropy? It has already been noticed that the chronological development of the subject is not being followed.

Therefore,

$$\frac{q_1 - q_2}{q_1} = \frac{T_1 - T_2}{T_1} ; \frac{T_2}{T_1} = \frac{q_2}{q_1},$$

defining a scale of temperature by relating a ratio of two temperatures to a ratio of two quantities of heat, and so again independent of any particular substance or system.

$$\frac{q_1}{T_1} = \frac{q_2}{T_2} ; \frac{q_1}{T_1} - \frac{q_2}{T_2} = 0 ;$$

remembering that q_1, heat taken in at T_1, and q_2, heat rejected at T_2, have both been counted as positive in the preceding 'Carnot argument', this with heat absorbed uniformly positive, can be generalised to

$$\oint \frac{dq_{rev}}{T} = 0 ; \frac{dq_{rev}}{T} = dS,$$

partially *defining entropy*; T^{-1} appears as the integrating factor leading to the complete differential dS. Thus $\oint dS = 0$; $\Delta S = S_2 - S_1$; $\Delta S = q_{rev}/T$.

All this is basic physics and is essential to the definition of free energy, which can well be taken as the chemical-thermodynamic version of the second law.

If the search for the nub has succeeded, it is worth looking again at the quotations, early in this section, that drove the student to drink. If the first is exaggerated, it does admit difficulties, but these mainly stem from entanglement of lines of thought. If the second is true, there are good reasons other than dim wits. The third is true, and only a reminder is needed to bring out the interesting reason why.

As far as we know, the first law (taking in mc^2 as a form of energy) is strictly followed by all events on any scale, nuclear, atomic, molecular, without exception. Its validity is not confined to macroscopic systems. The validity of the second law *is* so confined, as proven by the existence of observable fluctuations in sufficiently small systems (see section 4.7.2). The formation of flickering, microscopic hot and cold spots in a system at uniform temperature seems an affront to the classical second law; on the other hand, the statistical basis of the law is that all microstates are impartially explored. This is a paradox that, understood, pleases.

5.4   The four fundamental equations of Gibbs

All the working thermodynamic relationships we need flow directly from four fundamental equations, derived as follows.

Any infinitesimal process undergone by a system in equilibrium is a reversible process because, in the limit, it causes no departure from the state of equilibrium. For such an infinitesimal process, the 'first law' can be written (recognising $q_{rev} = TdS$) as

$$dE = TdS - PdV \tag{5.1}$$

This is the first of the four equations. The other three are obtained by first writing down the defining equations for enthalpy, Helmholtz free energy, and Gibbs free energy. These equations are differentiated generally, and the results are simplified by substitution for dE in terms of equation (5.1). Thus: $H = E + PV$; $dH = dE + PdV + VdP$

Therefore $dH = TdS + VdP$ (5.2)

$A = E - TS$; $dA = dE - Tds - SdT$

Therefore $dA = - SdT - PdV$ (5.3)

$G = E + PV - TS$; $dG = dE + PdV + VdP - TdS - SdT$

Therefore $dG = - SdT + VdP$ (5.4)

These equations, (5.1) to (5.4), so easily kept in mind or derived on the back of an envelope at any time, apply to all closed systems in equilibrium states. They incorporate the first and second laws, and, as they stand, constitute criteria of equilibrium. A usable criterion, or test, however, calls for attention to be confined to the influence of an infinitesimal test process on one variable at a time, the others being kept constant. For instance, using equation (5.1), S and V can be kept constant, so that dS and dV are zero. Then, for any infinitesimal test process, the existence of equilibrium requires $(dE)_{S,V} = 0$, representing that energy must be at a minimum. Alternatively, E and V can be kept constant; this leads to the requirement that $(dS)_{E,V} = 0$, representing that the entropy must be at a maximum. The latter criterion will be seen to correspond to the Clausius over-concise statement of the second law. In a sense, this puts the statement in proper perspective, because there are in all *twelve* criteria to be written down by inspection of the four fundamental equations. All are equally general and valid, and a choice can be made according to suitability and convenience. In chemical thermodynamics we mostly use equation (5.4), keeping T and P constant, so that the criterion of equilibrium is $(dG)_{T,P} = 0$, i.e., a minimum in Gibbs free energy.*

* It should be noted that $dX = 0$ is consistent with either maximum, minimum, or inflection in X as a function of an independent variable. The choice does not, of course, arise in the present context. At equilibrium, S and P tend to maxima, all the other functions to minima.

The first two equations have already been used to write down expressions for the temperature coefficients of entropy (Section 4.7.1). Occasion may be taken to derive the volume and pressure coefficients.

Using equation (5.3), remembering that dA is a complete differential, we get by cross-differentiation that

$$\left(\frac{\partial S}{\partial V}\right)_T = \left(\frac{\partial P}{\partial T}\right)_V \tag{5.5}$$

which can be immediately put to useful purposes. A glance at the right-hand side, however, shows a quantity not always easily measurable by a direct method. The difficulty is easily obviated by a typical manipulation. For any system, the volume is a function of temperature and pressure, i.e., $V = V(T, P)$, so that

$$dV = \left(\frac{\partial V}{\partial T}\right)_P dT + \left(\frac{\partial V}{\partial P}\right)_T dP$$

Imposing constancy of volume, $dV = 0$, so that

$$\left(\frac{\partial P}{\partial T}\right)_V = -\left(\frac{\partial V}{\partial T}\right)_P \Big/ \left(\frac{\partial V}{\partial P}\right)_T = \frac{\alpha}{\beta},$$

where α = coefficient of expansion = $\dfrac{1}{V}\left(\dfrac{\partial V}{\partial T}\right)_P$

and β = coefficient of compressibility = $-\dfrac{1}{V}\left(\dfrac{\partial V}{\partial P}\right)_T$

both directly measurable. Hence,

$$\left(\frac{\partial S}{\partial V}\right)_T = \frac{\alpha}{\beta} \tag{5.6}$$

Equation (5.5) can also be used to put $(C_P - C_V)$ in order. From equation (5.1), by inspection,

$$\left(\frac{\partial E}{\partial V}\right)_T = T\left(\frac{\partial S}{\partial V}\right)_T - P,$$

so that, using (5.5)

$$\left(\frac{\partial E}{\partial V}\right)_T = T\left(\frac{\partial P}{\partial T}\right)_V - P, \text{ or } P = T\left(\frac{\partial P}{\partial T}\right)_V - \left(\frac{\partial E}{\partial V}\right)_T \tag{5.7}$$

which is sometimes called, for some reason, a 'thermodynamic equation of state'. Recalling the previously derived relation between C_P and C_V (Section 4.3), namely

$$C_P - C_V = \left(\frac{\partial V}{\partial T}\right)_P \left[P + \left(\frac{\partial E}{\partial V}\right)_T \right]$$

it is seen that substitution for P in terms of equation (5.7) gives

$$C_P - C_V = T \left(\frac{\partial P}{\partial T}\right)_V \left(\frac{\partial V}{\partial T}\right)_P$$

Since (above)

$$\left(\frac{\partial P}{\partial T}\right)_V = \frac{\alpha}{\beta} \quad \text{and} \quad \left(\frac{\partial V}{\partial T}\right)_P = \alpha V$$

$$C_P - C_V = \frac{\alpha^2 V T}{\beta} \tag{5.8}$$

previously noted without derivation.

Equation (5.5) leads directly to the Clapeyron–Clausius equation. Consider a pure liquid in equilibrium with its own vapour at temperature T; the vapour pressure is constant, fixed only by temperature, independently of the relative amounts of the two phases present in the system (recollect the horizontal sections of the Andrews $P(V)$ isothermals, and, later, apply the Gibbs phase rule in retrospect). The vapour pressure is therefore independent of the total volume, and its temperature coefficient is also independent of volume, and can be written accordingly as $(\partial P/\partial T)_V$.* Then, using equation (5.5), and, conventionally, dropping the partial notation, $dP/dT = (\partial S/\partial V)_T$. But, for evaporation of a liquid at constant temperature, entropy increases linearly with total volume, so that $(\partial S/\partial V)_T$ is constant, and can be equated to $\Delta S/\Delta V$ for the whole process. Since, per mole, $\Delta S = \lambda/T$, where λ is the molar latent heat of evaporation,

$$\frac{dP}{dT} = \frac{\lambda}{T \Delta V} \tag{5.9}$$

which is the Clapeyron–Clausius equation, equally valid for any phase transition, or, indeed, for any equilibrium system with pressure solely dependent on temperature. Adaptations and applications of this equation— a rich field of numerical problems for the confusion of students—are not on our agenda.

* $(\ )_V$ is usually translated as 'at constant volume'. Total independence of volume is an equally valid translation when appropriate.

The entropy/pressure coefficient can be written down by inspection of equation (5.4); cross-differentiation gives

$$\left(\frac{\partial S}{\partial P}\right)_T = -\left(\frac{\partial V}{\partial T}\right)_P = -\alpha V \tag{5.10}$$

For the special case of a mole of ideal gas, it is seen that $\alpha = R/PV$, so that $dS = (-R/P)dP$. Then, for a finite change of pressure at constant temperature,

$$\Delta S = -\int_{P_1}^{P_2} \frac{R}{P}\, dP = R \ln \frac{P_1}{P_2}, \tag{5.11}$$

a result obtained previously.

5.5   The three main working equations

Three equations of the chemical thermodynamics of closed systems are derived with singular ease; ability to derive, discuss and apply them represents significant competence in thermodynamics, adequate for a number of useful purposes—including the frustration of ill-intentioned examiners.

These equations all come from the fourth fundamental equation (equation 5.4), namely, $dG = -SdT + VdP$. Derived from this by inspection we already have $(dG)_{T,P} = 0$, and now write down

$$\left(\frac{\partial G}{\partial T}\right)_P = -S \tag{5.12}$$

and

$$\left(\frac{\partial G}{\partial P}\right)_T = V \tag{5.13}$$

Not very much to be remembered?

5.5.1   The Gibbs–Helmholtz equation

Since G and S are extensive properties, so that for a change from state 1 to state 2, $\Delta G = G_2 - G_1$ and $\Delta S = S_2 - S_1$, equation 5.12 leads to

$$\left(\frac{\partial \Delta G}{\partial T}\right)_P = -\Delta S \tag{5.14}$$

or, if we wish to think in terms of affinity,

$$\left\{\frac{\partial(-\Delta G^\circ)}{\partial T}\right\}_P = \Delta S^\circ \tag{5.15}$$

which tells us that a reaction accompanied by an increase in the entropy *of the system in which the reaction takes place* is thermodynamically favoured by rise of temperature, as previously discussed.

Since for a reaction at constant temperature and pressure $\Delta H = \Delta G + T\Delta S$, or $\Delta S = (\Delta H - \Delta G)/T$, equation (5.14) may be written

$$\left(\frac{\partial \Delta G}{\partial T}\right)_P = \frac{\Delta G - \Delta H}{T}$$

or $$\Delta H = \Delta G - T\left(\frac{\partial \Delta G}{\partial T}\right)_P \tag{5.16}$$

which are alternative forms of the Gibbs–Helmholtz equation. It furnishes a non-calorimetric method of determining enthalpy change. For the purpose of expressing ΔG as a function of temperature, it is conveniently expressed*

$$\left\{\frac{\partial\left(\dfrac{\Delta G}{T}\right)}{\partial T}\right\}_P = -\frac{\Delta H}{T^2} \tag{5.17}$$

5.5.2   The reaction isotherm

For simplicity, let us consider a formalised 'ideal gas reaction'

$$A + B = 2C$$

The relation to be studied is that between K_P for this reaction and ΔG°, the standard Gibbs free energy change that 'accompanies it'. To define the latter, standard states must be specified. They are the pure, participating gases A, B and C, all at temperature T, and at arbitrarily chosen standard pressures, P_A°, P_B° and P_C°, which may be the same or different. The standard Gibbs free energy change is therefore to be regarded as

$$\Delta G^\circ = \Sigma n G^\circ = 2G_C^\circ - G_A^\circ - G_B^\circ \tag{5.18}$$

$$*\left\{\frac{\partial\left(\dfrac{\Delta G}{T}\right)}{\partial T}\right\}_P = \frac{1}{T}\left(\frac{\partial \Delta G}{\partial T}\right)_P - \frac{\Delta G}{T^2} = \frac{\Delta G - \Delta H}{T^2} - \frac{\Delta G}{T^2} = -\frac{\Delta H}{T^2}$$

where the G° terms represent standard Gibbs free energies per mole of the substances concerned in their standard states, which must always be specified.

For any mixture of A, B and C, at temperature T, in which the chemical equilibrium has been established, let the partial pressures and Gibbs free energies per mole of the participants be denoted P_A^e, P_B^e, P_C^e and G_A^e, G_B^e, G_C^e. To confirm the existence of equilibrium, let us use the criterion $(dG)_{T,P} = 0$. This amounts to pushing the reaction, at constant T, P, infinitesimally in either direction; if $(dG)_{T,P} >$ or < 0 equilibrium has not been attained. If $(dG)_{T,P} = 0$, it has been. In the latter case,

$$2\,dn_C G_C^e - dn_A\,G_A^e - dn_B\,G_B^e = 0$$

Since the infinitesimal fractions of moles must satisfy stoichiometry, $dn_A = dn_B = dn_C$, so that

$$\Delta G^e = 2G_C^e - G_A^e - G_B^e = 0 \tag{5.19}$$

i.e., we have a blinding glimpse of the obvious—the transformation of reactants in equilibrium with products into products in equilibrium with reactants causes no free energy change.

A relation between the G^e and G° terms is required. Why do these terms differ? Because the P^e and P° terms differ. For each ideal gas participant, the enthalpy is independent of pressure at constant temperature, but the entropy is a function of pressure (partial or total) as in equation (5.10) or (5.11). Consequently the Gibbs free energy is a function of pressure, as, indeed is written in equation (5.13). Using this,

$$G_A^e = G_A^\circ + \int_{P_A^\circ}^{P_A^e} V\,dP = G_A^\circ + RT\ln\frac{P_A^e}{P_A^\circ} \tag{5.20}$$

and similarly for B and C. Substitution in equation (5.19) gives

$$2G_C^\circ + 2RT\ln\frac{P_C^e}{P_C^\circ} - G_A^\circ - RT\ln\frac{P_A^e}{P_A^\circ} - G_B^\circ - RT\ln\frac{P_B^e}{P_B^\circ} = 0$$

and on rearrangement,

$$\underbrace{2G_C^\circ - G_A^\circ - G_B^\circ}_{\Delta G^\circ} + \underbrace{RT\ln\frac{(P_C^e)^2}{P_A^e P_B^e}}_{RT\ln K_P} - RT\ln\frac{(P_C^\circ)^2}{P_A^\circ P_B^\circ} = 0$$

Therefore,

$$\Delta G^\circ = - RT \ln K_P + RT \ln \frac{(P_C^\circ)^2}{P_A^\circ P_B^\circ} \qquad (5.21)$$

which is the reaction isotherm for this formal reaction.

Since, in this equation, all terms other than K_P are constants (natural, agreed, or functions of fixed standard states), K_P itself must be a constant term. This is the traditional thermodynamic 'proof' of the law of mass action.

If all the standard pressures are set at 1 atm (the P^e terms also being expressed in these units), the second term on the right-hand side of equation (5.21) comes to zero. The reaction isotherm is indeed often expressed as

$$\Delta G^\circ = - RT \ln K_P \qquad (5.22)$$

but this imposes a special meaning on ΔG°, sometimes on K_P as well, and sets some pitfalls to be avoided. Looking at the last equation, the thought comes up that not every reaction (like the formal example) is isomolecular. If it is not, K_P has dimensions and must be expressed in appropriate units —atm^2, atm^{-1}, whatever. What is the significance of the logarithm of anything but a dimensionless number? It is a nonsense. Here we can take a leaf out of recent recommendations (already sufficiently publicised) as follows. Suppose that, for a given equilibrium (perhaps a dissociation) it has been found that $K_P = 0\cdot5$ atm. This result can equally well be stated as $K_P/\text{atm} = 0\cdot5$, and should be, but hardly ever is. In other words, the numerical value of any measured quantity is to be regarded as the ratio of the quantity and the unit used in its measurement, and we can take the log of this quite happily. It could be argued that this is, in effect, what happens when we set out an equation like equation (5.21), provided that all the P° terms have been set equal to one unit of whatever pressure scale is being used. Very well, but the second term on the right-hand side is therefore important—if it vanishes numerically, it may retain its dimensional significance. There is another reason why this term remains essential. It is that this particular choice of the P° values is frequently inappropriate, as will be seen. It is therefore clear that use of equation (5.22) is hedged around with rules and conditions, to be forgotten at peril of muddle. A reasonable compromise is to remember the reaction isotherm in the form

$$\Delta G^\circ = - RT (\ln K_P - \Sigma n \ln P^\circ) \qquad (5.23)$$

Applications of the isotherm are best illustrated by specific examples.

Bodenstein and Starck, in 1910, studied the equilibrium $I_2 \rightleftharpoons 2I$ at high temperatures, enclosing the gaseous system in a totally sealed, all-silica vessel, the pressure inside being measured by their celebrated 'quartz spring manometer' (working on the principle of the Bourdon gauge). At 1000 °C, they found $K_P = 0 \cdot 165$ atm (note, $K_C = 1 \cdot 58 \times 10^{-3}$ mole 1^{-1}), identical with the more recent spectroscopically determined value.[3] Problem: what is $\Delta G°$ for this dissociation at 1000 °C? Using equation (5.23):

$$I_2(g, 1\,atm) = 2I(g, 1\,atm);$$

$$\Delta G° = -1 \cdot 987 \times 1273 \times 2 \cdot 303\,(\log K_P - \Sigma n \log P°)$$

$$= -5825\,(\log 0 \cdot 165 - 0)$$

$$= -5825. \times \bar{1} \cdot 2178 = 4552\ cal.$$

$\Delta G°$ is positive, indicating not that dissociation does not occur, only that it is not extensive. There is no reason why we should not vary the standard states—maintaining the same temperature, but altering the standard pressures to anything we please. Let us therefore examine, however foolishly,

$$I_2(g, 1\,atm) = 2I(g, 0 \cdot 1\,atm);$$

$$\Delta G° = -5825\left(\log 0 \cdot 165 - \log \frac{(0 \cdot 1)^2}{1}\right)$$

$$= 4552 - 5825 \times 2 = -7093\ cal.$$

$\Delta G°$ is now negative; this is because of the increased entropy of the product. Are we justified in using the same symbol, $\Delta G°$, for two different quantities like this? There might be disagreement here, but the writer thinks it is right, because in each case $\Delta G°$ is a quantity depending in value on arbitrary fixing of standard states. It, with other symbols like it, is never to be translated without enquiry as to *what* standard states are being used to give it quantitative significance.

As yet another (*reductio ad absurdum*) illustration of this facility, let us examine

$$I_2(g, 0 \cdot 668\,atm) = 2I(g, 0 \cdot 332\,atm);$$

$$\Delta G° = -5825\left(\log 0 \cdot 165 - \log \frac{(0 \cdot 332)^2}{0 \cdot 668}\right)$$

$$= 4552 - 4552 = 0.$$

$\Delta G°$ is zero. This is because, with great futility, the standard pressures have been adjusted to equality with the equilibrium partial pressures. A ludicrous exercise perhaps, but a reminder that $\Delta G^e = 0$, and that we need not be inhibited in choosing 'standard' states—the isotherm gives the appropriate answer, whatever its degree of usefulness.

The value of the reaction isotherm of course depends on its extension to equilibria involving condensed phases. *How* can be shown by another 'problem': If the dissociation pressure of nickel oxide is $5\cdot181 \times 10^{-26}$ mm at 450 °C, what is $\Delta G°$ for the formation of nickel oxide at this temperature?

Some preliminary comments are needed. Quotation of the dissociation pressure of a substance implies that it is in equilibrium with its dissociation products, which must therefore be present. Using the Gibbs phase rule (in advance of presenting it), the Ni, NiO, O_2 system is seen to be a two-component, three-phase system, and is therefore ($F = C - P + 2$) uni-variant, i.e., the equilibrium pressure depends only on the temperature. The Clapeyron–Clausius, or some allied equation, must therefore apply, as considered in the following section. The given dissociation pressure (understood to be of oxygen) is immeasurably small—such as would be exerted by one molecule in several million litres. It may be asked, how can such a quantity be physically significant, or meaningfully different from zero? If it is indeed significant, would it not be better expressed in some other way?

The problem is, of course, artificial in this respect—the pressure is not directly measured by any pressure-sensing device—but perhaps serves a purpose in suggesting the need for an intensive property, capable of quantitative expression, of the nature of the '*escaping tendency*', conceived by G. N. Lewis in 1900. This helpful idea is self-explanatory. It is meaningful to say, for example, that at equilibrium the escaping tendencies of 'nickel substance' from metallic nickel and from nickel oxide must be identical, or that the escaping tendency of oxygen from *nickel–nickel oxide couple* at 450 °C (by no means insignificant) must be the same as that of oxygen from molecular oxygen gas at the pressure quoted. Were this not so, there would be disequilibrium, and a natural transfer of one or the other component from the situation of higher to one of lower escaping tendency until equalisation was achieved and a state of equilibrium attained. How this concept is quantified is a major topic of the following chapter; in the meantime, pressure serves well, no matter of what order of magnitude.*

The formation reaction of nickel oxide can be considered as proceeding, at equilibrium, in the gas phase (Fig. 5.3.). Crystalline nickel has, in principle, a vapour pressure, and can be transferred to the gas state, at a pressure P_{Ni}^e defined by equilibrium with the solid nickel, with zero free energy change. Similarly, the transfer of nickel oxide between its solid

---

* Consider the difference between zero, a definite number, and the indefinite smallness of a quantity, initially real without doubt, progressively decreased through the negative powers of ten: 10^{-1}, 10^{-2}... 10^{-10}... 10^{-100}.... Where is 'significance' (absolute, not relative) to be cut off?

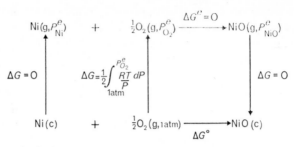

Fig. 5.3   Hypothetical gas phase formation of nickel oxide

state and the gaseous state in equilibrium with it, at vapour pressure P^e_{NiO}, involves $\Delta G = 0$. The oxygen can be reversibly changed in pressure from 1 atm to the pressure $P^e_{O_2}$ defined by the chemical equilibrium with the gaseous nickel and the gaseous nickel oxide. For the gaseous formation reaction proceeding under equilibrium conditions, $\Delta G^e = 0$. It can be seen from Fig. 5.3 that ΔG° for the normal formation reaction is

$$\Delta G^\circ = \tfrac{1}{2} RT \ln P^e_{O_2} = RT \ln (P^e_{O_2})^{\frac{1}{2}}$$

with $P^e_{O_2}$ expressed in atm.

The reaction isotherm offers a more concise argument. For the *dissociation*, we start writing it out as follows:

$$\Delta G^\circ = -RT \left\{ \ln \frac{P^e_{Ni}(P^e_{O_2})^{\frac{1}{2}}}{P^e_{NiO}} - \frac{?\,(P^o_{O_2})^{\frac{1}{2}}}{?} \right\}$$

but we get stuck, as indicated by the question marks. What entries should be made representing the standard states of nickel and nickel oxide? There is, of course, only one rational answer. The states of these solid phases are not open to arbitrary adjustment; the only states are standard and equilibrium states at the same time, so we have to write in P^e_{Ni} and P^e_{NiO} in the appropriate gaps. Cancellation of these imponderable quantities occurs, leaving

$$\Delta G^\circ = -RT \ln (P^e_{O_2})^{\frac{1}{2}}/(P^o_{O_2})^{\frac{1}{2}} = -RT \ln K_P$$

for the dissociation. The first equality, with $P^o_{O_2} = 1$ atm, and appropriate reversal of sign, agrees with the previous result; the second indicates how K_P is *defined* in such cases.

Proceeding now with the calculation,

$$\Delta G^\circ_f = \tfrac{1}{2} \times 1 \cdot 987 \times 723 \times 2 \cdot 303 \log \frac{5 \cdot 181 \times 10^{-26}}{760}$$

$$= -46 \cdot 6 \text{ kcal, not of surprising magnitude.}$$

It might be asked, what if all reactants and products are solids, so that all the equilibrium pressures are individually fixed; how is the chemical equilibrium then to be satisfied? Consider, for example, $Fe(s) + S(s) = FeS(s)$?

The answer is that this would be a two-component, four-phase (including vapour) system, and therefore invariant, i.e., all the equilibria could be satisfied only at a unique quadruple point, fixed in temperature and pressure. At any other temperature, one of the solid phases must disappear, releasing one partial pressure to assume a less-than-saturation value, self-adjusting to suit the chemical equilibrium. The situation then becomes the same as in the above example.

The importance of the reaction isotherm is twofold. Applied to experimental equilibrium studies, it supplements the store of $\Delta G°$ data. On the other hand, using $\Delta G°$ data from the store, it enables K_P values to be calculated.

5.5.3   The van't Hoff equation

This is the third of the working equations. Its relation to the other two is apparent if all three are set out in triangular display:

Gibbs-Helmholtz

$$\left\{ \frac{\partial}{\partial T}\left(\frac{\Delta G°}{T} \right) \right\}_P = -\frac{\Delta H°}{T^2}$$

Reaction Isotherm

$$\frac{\Delta G°}{T} = -R \ln K_P$$

---

van't Hoff

$$\left(\frac{\partial \ln K_P}{\partial T} \right)_P = \frac{\Delta H°}{RT^2}$$

(5.24)

It seems that the three equations convey but two messages, and that either $\Delta G°$ or $\ln K_P$ could equally well be used in thought and calculation. In practice, these functions do have their own particular affiliations— perhaps as much by tradition as otherwise—with particular experimental fields and applications. The Gibbs–Helmholtz (in the form of equation (5.16)) is classically associated with electrochemical determination of $\Delta G°$ and its temperature coefficient. It is the basis of the practical expression of

ΔG° data, such as are stored, as functions of temperature over a wide range. On the other hand, the van't Hoff equation is directly applicable to equilibrium studies, involving either gas or solution phases. Both can be listed as providing routes, alternative to calorimetry, to ΔH° data— sometimes more accurate, sometimes obligatory (e.g., ΔH° for $H_2S \rightleftharpoons H_2 + \frac{1}{2}S_2$ at high temperatures). It is of interest to compare the van't Hoff equation with the Clapeyron–Clausius equation for liquid–vapour equilibrium, and to consider the functional relationship between them. Temperature dependence is the business of both Gibbs–Helmholtz and van't Hoff equations, but each is suited to particular aspects, as will be shown.

Since ΔG° is directly available from $\ln K_P$, solution of the van't Hoff equation for ΔH° gives all three terms in $\Delta G^\circ = \Delta H^\circ - T\Delta S^\circ$; some inkling of the chemical interest of this has already been given. For a given equilibrium, two equilibrium constants at different temperatures comprise the minimum information needed to solve the (differential) equation, but this is clearly a rough and ready expedient. A better application is to use the equation in integrated form.

$$\ln K_P = -\frac{\Delta H^\circ}{T} + \text{const} \tag{5.25}$$

and to plot values of $\ln K_P$, determined over a range of temperatures, against $1/T$. The plot can be examined, and, if it is rectilinear within experimental error (as often the case), the slope, $-\Delta H^\circ/R$, provides an average value of ΔH° over the experimental temperature range. In so far as ΔH° is only coincidentally independent of temperature (this would require $\Delta C_P^\circ = 0$), such a plot cannot be truly rectilinear, and equation (5.25) can be used in this way only as an approximation, sometimes adequate, but not always—the problem is still being nibbled at. A fresh approach is needed if all the information in principle available from a (hypothetical) set of $\ln K_P(T)$ measurements of unlimited accuracy is to be extracted.

Experimental* $\ln K_P$ values, adequate in number, spread over a wide enough temperature range, must be expressed as a function of temperature by means of an equation (in the simplest case, an empirical polynomial in T) fitted by a satisfactory statistical method (e.g. the method of least squares[5]). Differentiation with respect to temperature gives an equation for ΔH° as a function of temperature. This enables the terms of $\Delta G^\circ = \Delta H^\circ - T\Delta S^\circ$ to be obtained at precise temperatures within the range studied. Considerable interest, especially in the field of equilibria in solution, attaches to ΔC_P°. Yet another differentiation with respect to temperature

---

* K_P values never are strictly 'experimental'; masses, volumes, temperatures, all sorts of physical quantities, all subject to experimental errors, go into deriving them.

will give a third equation—for ΔC_P° (T). Unfortunately, such a sequence of calculations is subject to a heavy disability, *well to be remembered as a matter of general principle*, indeed, perilous to forget. It is that each successive differentiation of a function with respect to an independent variable is attended by loss of significance. The effect of experimental error progressively increases, and one or two 'significant figures' must be chopped off the data with each differentiation. It then becomes obligatory to apply tests of significance[5] (in the light of the 'internal error of the experiment') to each advance into the unknown. Computer programmes exist for alternative methods[6] of tackling this important, if specialised, project of deriving thermodynamic functions from equilibrium constants. It is a relevant concluding comment that there is particular point, in such circumstances, in striving to improve the accuracy of physical measurements to limits that otherwise might be judged absurd.

There are two aspects of the temperature dependence of ΔG° to be considered. The first could be called general and fundamental, starting with the familiar $\Delta G^\circ = \Delta H^\circ - T\Delta S^\circ$ for a reaction at constant temperature and pressure. How can the influence of change of temperature on the reaction be analysed, and how can one reaction be compared with another in this respect?

We know that ΔH° varies with temperature as described by the Kirchhoff theorem, also that $\Delta S^\circ = \Sigma nS^\circ$ and that $(\partial S^\circ/\partial T)_P = C_P^\circ/T$. We can therefore expand the right-hand side of the primary equation to obtain

$$\Delta G^\circ = \Delta H_0^\circ = \int_0^T \Delta C_P^\circ \, dT - T\int_0^T \frac{\Delta C_P^\circ}{T} \, dT - T\Delta S_0^\circ \qquad (5.26)$$

for the whole range of temperature above the absolute zero. Phase transitions will, of course, contribute $\Sigma n\lambda_{tr}$ and $\Sigma n\lambda_{tr}/T_{tr}$ terms. If reactants and products do not deviate from the third law, $\Delta S_0^\circ = 0$. In any event, at $T = 0$, $\Delta G_0^\circ = \Delta H_0^\circ = \Delta E_0^\circ$. With temperature rising from 0 °K, the other two terms become increasingly significant. It is interesting to observe the dominant role of heat capacity. This equation bears lengthy contemplation; it summarises a great deal of earlier discussion, and could be said to be the nub of the thermodynamics of closed systems.

The second aspect is practical: how ΔG° data over a wide temperature range are handled.

Integration of the Gibbs–Helmholtz equation, recognising the temperature dependence of ΔH°, gives

$$\frac{\Delta G^\circ}{T} = -\int \frac{\Delta H^\circ}{T^2} \, dT + J \qquad (5.27)$$

where J is an integration constant. Kirchhoff deals with the variation of ΔH° with temperature (cf. section 3.5), in terms of the $C_P^{\circ}(T)$ equations for all reactants and products, i.e.,

$$\Delta H^{\circ} = \Delta H_0^{\circ} + \Delta\alpha T + \frac{\Delta\beta}{2} T^2 + \frac{\Delta\gamma}{3} T^3 \ldots \tag{5.28a}$$

Substitution in (5.27) gives

$$\frac{\Delta G^{\circ}}{T} = -\int \left(\frac{\Delta H_0^{\circ}}{T^2} + \frac{\Delta\alpha}{T} + \frac{\Delta\beta}{2} + \frac{\Delta\gamma}{3} T \ldots \right) dT + J$$

$$= \frac{\Delta H_0^{\circ}}{T} - \Delta\alpha \ln T - \frac{\Delta\beta}{2} T - \frac{\Delta\gamma}{6} T^2 \ldots + J$$

Therefore,

$$\Delta G_0^{\circ} = \Delta H_0^{\circ} - \Delta\alpha T \ln T - \frac{\Delta\beta}{2} T^2 - \frac{\Delta\gamma}{6} T^3 \ldots + JT \tag{5.29a}$$

where, it is to be remembered, ΔH_0° normally has no more significance than that of an integration constant—*unless* the heat capacity measurements are of 'third law type' and extend to temperatures not much above $0\,°K$. For putting numerical values to the two integration constants, one value each of ΔH° and ΔG° at specified temperatures (the same or different) are required.

The first part of such a calculation (to equation (5.28a)) has been illustrated in section 3.5; it can now be extended, taking in the additional datum needed, i.e.,

$$H_2\,(g, 1\,atm) + \tfrac{1}{2}O_2\,(g, 1\,atm) = H_2O\,(g, 1\,atm); \quad \Delta G_{298\cdot15}^{\circ} = -54635\cdot7\,cal$$

This, with the previous figures, leads to the quantified version of equation (5.29a) for this reaction:

$$\Delta G^{\circ} = -57016\cdot4 + 6\cdot3805\,T \log_{10} T - 0\cdot4735 \times 10^{-3}\,T^2 -$$

$$- 0\cdot4917 \times 10^{-7}\,T^3 - 7\cdot6575\,T\,cal \tag{5.30a}$$

valid between $298\cdot15$ and $1500\,°K$.

If the alternative expression of molar $C_P^{\circ}(T)$ has been used (section 3.5 $C_P^{\circ} = a + bT + cT^{-2}$), parallel development leads to

$$\Delta H^{\circ} = \Delta H_0^{\circ} + \Delta aT + \frac{\Delta b}{2} T^2 - \Delta cT^{-1} \tag{5.28b}$$

and to

$$\Delta G^{\circ} = \Delta H_0^{\circ} - \Delta aT \ln T - \frac{\Delta b}{2} T^2 - \frac{\Delta c}{2} T^{-1} + IT \tag{5.29b}$$

which, consistently with the figures used in section (3.5), becomes for the reaction in question

$$\Delta G^\circ = -56992 + 6\cdot4472\, T \log_{10} T - 0\cdot59 \times 10^{-3} T^2 -$$
$$- 0\cdot04 \times 10^{-5} T^{-1} - 7\cdot8403T \qquad (5.30b)$$

Since $\Delta G^\circ/T = -R \ln K_P$, alternative equations can be written for $-R \ln K_P$ as a function of temperature; these are evidently integrated van't Hoff equations, and lend themselves better to fields where the experimental method is the direct measurement of equilibrium states. For instance, equation (5.29b) leads to

$$-R \ln K_P = \frac{\Delta H_0^\circ}{T} - \Delta a \ln T - \frac{\Delta b}{2} T - \frac{\Delta c}{2} T^{-2} + I \qquad (5.31)$$

On rearranging

$$-R \ln K_P + \Delta a \ln T + \frac{\Delta b}{2} T + \frac{\Delta c}{2} T^{-2} = \frac{\Delta H_0^\circ}{T} + I = \sum \quad (5.32)$$

Then a so-called 'sigma plot'; of the left-hand side of this equation against $1/T$, should be rectilinear, with slope ΔH_0°. and intercept equal to the integration constant I. This provides the best method of securing the most representative values of the two unknowns, ΔH_0° and I, from equilibrium constants determined over a wide range of temperatures. The same procedure can equally well be based on equation (5.29a); Fig. 5.4 is a sigma plot for the $H_2 + \frac{1}{2}O_2 \rightleftharpoons H_2O$ equilibrium, incorporating the information

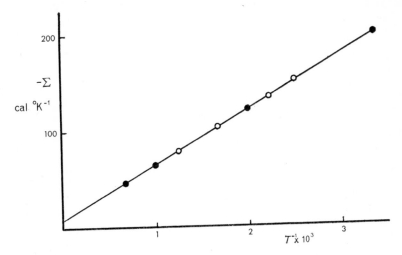

Fig. 5.4 Sigma plot for $H_2 + \frac{1}{2}O_2 \rightleftharpoons H_2O$

contained in both equations (5.30a) and (5.30b). The method, clearly the best basis for cautious extrapolation, is used in metallurgy for the vapour pressures of metals at high temperatures—the basic equation is then, of course, the Clapeyron–Clausius equation.

5.6 The storage and retrieval of standard data

The importance of continued critical selection and central tabulation of standard enthalpies and Gibbs free energies of formation of compounds needs no emphasis. There is a tremendous gain in usefulness if such data can be released from their initial restriction to 25 °C and extended to a wide operational temperature range. Equations (5.28) and (5.29) can be, and are, used for this purpose. Applied to formation reactions, such equations (all containing like functions of T) can be used in term-by-term algebraic summations (based of course on the status of G and H as extensive properties) to arrive at equations appropriate to any reaction it is desired to study.

Let us consider what is the minimum of quantitative information that must be tabulated to allow calculation of $\Delta H°$ and $\Delta G°$ for a given formation reaction at any temperature. $\Delta H°_{298}$ and $\Delta G°_{298}$ are already, traditionally, tabulated; $\Delta\alpha$, $\Delta\beta$ and $\Delta\gamma$ are also required, so that five entries per compound would normally be adequate (without allowing for phase changes). Since each element forms many compounds, however, it might be better to tabulate the individual heat capacity coefficients, α, β and γ, for each substance, element or compound, separately. In this case, the worker, having looked up the tables, has to perform a calculation such as already illustrated.

Another way of presenting such information has come into use, for a number of reasons. The first is that, as indicated, equations like (5.30)—one for each reaction—are a little cumbersome. Plots of $\Delta G°$ against T, although often nearly straight, vary widely in slope. It would be nice to read off what one needs from tabulations, at equally spaced temperatures, of some function changing but slowly with temperature, making interpolation easy. There are more basic reasons, as follows.

All $\Delta X°$ values are differences. There would be advantage in a system based only on $X°$—extensive properties of substances, elements equally with compounds. As it stands, the traditional $\Delta H°$ and $\Delta G°$ of formation of compounds are, in effect, based on $H° = G° = 0$ for all elements in their standard states. Such an arbitrary assignment (not permissible for $S°$) can be made only at *one* temperature, and the choice of 25 °C is arbitrary and irrational.

Suppose, therefore, a new start is made, on the basis that the only temperature with any absolute significance is 0 °K. For each pure substance, element or compound, we write

$$G_T^\circ = H_T^\circ - TS_T^\circ$$

for the extensive properties per mole concerned, at the finite temperature T — never mind whether or not they can be expressed on an absolute scale. Since

$$H_T^\circ = H_0^\circ + \int_0^T C_P^\circ \, dT \quad \text{and} \quad S_T^\circ = \int_0^T \frac{C_P^\circ}{T} \, dT$$

then, ignoring any failures of the third law,

$$G_T^\circ = H_0^\circ + \int_0^T C_P^\circ \, dT - T \int_0^T \frac{C_P^\circ}{T} \, dT \tag{5.33}$$

This is an equation for a *substance*, to be compared with equation (5.26) for a *reaction*. Rearrangement gives

$$-\left(\frac{G_T^\circ - H_0^\circ}{T}\right) = \int_0^T \frac{C_P^\circ}{T} \, dT - \frac{1}{T} \int_0^T C_P^\circ \, dT \tag{5.34}$$

which defines the so-called *free energy function*. It has the properties desired for tabulation, and is seen to contain all the consequences of raising the temperature from 0 °K to T °K. It is absolutely determinable, either by statistical mechanical methods (for gases only), or from heat capacities measured to appropriately low temperatures.

To determine ΔG_T° for a reaction, the equation

$$\Delta G_T^\circ = T\Delta\left(\frac{G_T^\circ - H_0^\circ}{T}\right) + \Delta H_0^\circ \tag{5.35}$$

is used. This requires tabulation of ΔH_0°, derived from

$$\Delta H_0^\circ = \Delta H_{298}^\circ - \Delta(H_{298}^\circ - H_0^\circ) \tag{5.36}$$

The term in brackets is again obtained by statistical mechanical calculation or by reference to determined $C_P^\circ(T)$.

The alternative methods are illustrated in Table 5.1.

It is obvious that this scheme is not universally practicable. In any case, when the main interest is in chemical reactions at temperatures above 25 °C,

TABLE 5.1 Calculations of ΔG_T° for H_2 (g, 1 atm) + $\frac{1}{2}O_2$ (g, 1 atm) = H_2O (g, 1 atm)

(1) By solution of equation (5.30)				
T, °K	298·15	500	1000	1050
ΔG_T°, kcal	−54·64	−52·36	−46·06	−39·34

(2) By use of free energy functions[7]

	$\left(\dfrac{G_T^\circ - H_0^\circ}{T}\right)$, cal °K^{-1}				H_0°, cal
T, °K	298·15	500	1000	1500	
H_2O (g)	37·17	41·29	47·01	50·60	−57 107
H_2	24·42	27·59	32·74	35·59	0
O_2	42·06	45·68	50·70	53·81	0
$H_2O - (H_2 + \frac{1}{2}O_2)$	−8·28	−9·50	−11·08	−11·895	−57 107
$T\Delta\left(\dfrac{G_T^\circ - H_0^\circ}{T}\right)$	2469	4750	11 080	17 842	
ΔG_T°, kcal	−54·64	−52·36	−46·03	−39·27	

the profit of bringing in low temperature C_P° measurements seems questionable. This is no doubt why free energy functions based on H_{298}°, with ΔH_{298}°, are *also* tabulated; $-\left(\dfrac{G_T^\circ - H_{298}^\circ}{T}\right)$ at 298·15 °K is then S_{298}° in disguise. The procedure is the same, and the two systems are easily correlated. It seems to the writer that these formalisms, which add nothing in substance or principle, may be of advantage to the few habitually handling data, but otherwise they are an embarrassment. Ability to sort them out has, however, become essential for extracting information from standard works.

References

1. Ehrenberg, W., *Scient. American*, 1967, **217**, 103.
2. Paul, M. A., *Principles of Chemical Thermodynamics*, McGraw-Hill, 1951, pp. 202, 218.
3. Gibson, G. E., and Heitler, W., *Z. Physik*, 1928, **49**, 465.
4. Lewis, G. N., and Randall, M., *Thermodynamics*, McGraw-Hill, 1923, p. 176.
5. Bennett, C. A., and Franklin, N. L., *Statistical Analysis in Chemistry and Chemical Industry*, John Wiley, 1954.
6. Clarke, E. C. W., and Glew, D. N., *Trans. Faraday Soc.*, 1966, **62**, 539.
7. Pitzer, K. S., and Brewer, L., *Thermodynamics*, McGraw-Hill, 1961.

6

The Chemical Potential of a Component

6.1   Introduction

'Open systems' might have been a more orthodox title for this chapter, but, instead, place of honour has been given to the full name of a function to be regarded as a culmination of thermodynamic thought. 'Chemical potential', for short, would not do; despite colloquial usage, it has no meaning separated from the no less important concept of *component*.

Earlier discussion has presented Gibbs free energy as the quantity best adapted to thinking about equilibrium and affinity. This remains good, but can be developed along the following line of thought. Should not Gibbs free energy, in common with other forms of energy, have its own *intensity factor*, or potential? Temperature is heat potential. Heat flows down a temperature gradient; uniformity of temperature denotes thermal equilibrium. Transfer of charge occurs in response to electrical potential difference, and so on. The need for such a potential of Gibbs free energy, to quantify 'escaping tendency' has already been felt. But no such intensity factor can be defined independently of the *capacity factor* to go with it. This is where 'component' comes in. The word is familiar and well understood in a general context; this is the trouble—it will need specific definition in a particular context, requiring the chemist, especially, to clarify his ideas.

Apparently perversely, most of the chapter is devoted to *physical* equilibria, but—it is a good question—where is a line to be drawn between 'physical' and 'chemical' interactions? It will become clear enough that the chemical potential of a component is a unifying function valid for all kinds of equilibria and processes in closed and open systems. Yet, this demonstrated, it must be shown that other potentials sometimes play an essential part in determining 'the nature of things', and are not to be disregarded in viewing the scope of thermodynamic methods.

6.2 Some basic thermodynamics of mixtures and solutions

Since pure substances are not the rule, methods are needed to deal with multi-component, mixed systems, varying in composition. What consequences can be foreseen of mixing pure substances A and B together? Whereas, before mixing, all molecules of A and B had nearest neighbours of like kind, after mixing they have nearest neighbours of assorted kinds. If A and B are ideal gases, the results of this desegregation are minimal. Each gas exerts its own partial pressure independently of the presence of the other (Dalton's law), so that it is reasonable to consider the mixing process to be an ideal one. If, on the other hand, A and B are liquids, with fairly close-packed, strongly interacting molecules differing in shape and size, mixing (if it will occur) may have profound effects. What thermodynamic functions can deal, quite generally, with all possibilities? If such functions are of sufficiently general application, they will be suitable for discussion of pure systems as well.

6.2.1 Entropy of mixing

Let us consider first the ideal mixing of two gases, A and B, at constant temperature and pressure. Initially we have gas A at pressure P, and gas B at pressure P, perhaps in two compartments separated by a shutter. Pull up the shutter. Each gas diffuses spontaneously into the other, and, when mixing is complete, the partial pressures of A and B in the homogeneous mixture are P_A and P_B, both less than the unchanged total pressure, P. The process is quite irreversible, no work is done and no heat is absorbed. But, in principle, the natural process of mixing could be carried out reversibly, with w_{max} won, and q_{rev} absorbed. Whichever way the mixing is done makes no difference to the ΔX of mixing, where X is any extensive property. ΔX is determined only by final state (mixed) and initial state (unmixed). This agreed, we can calculate in terms of reversible mixing, and there is nothing more to it than the previously considered isothermal, reversible expansion of an ideal gas (section 4.4)—an exclusively 'entropy-driven' natural process. Each gas expands into the volume occupied by the other; E is constant, H is constant, but S is not.

Suppose that we mix n_A mole of A with n_B mole of B. The initial pressures are both P, and the final (partial) pressures are P_A and P_B. Adding the effects of the two expansions, in terms of the appropriate entropy for an ideal gas (section 5.4)

$$\Delta S_{mix} = n_A R \ln \frac{P}{P_A} + n_B R \ln \frac{P}{P_B}$$

But $\quad P_A = Pn_A/(n_A + n_B) = Px_A$

where x_A is the mole fraction of A in the mixture, and similarly for B. Therefore

$$\frac{P}{P_A} = \frac{1}{x_A}; \quad \frac{P}{P_B} = \frac{1}{x_B}$$

If we arrange that $n_A + n_B = 1$, then, for the formation of *one mole of mixture*,

$$\Delta S_{mix} = -x_A R \ln x_A - x_B R \ln x_B$$

and this can be generalised:

$$\Delta S_{mix} = -R \Sigma x_i \ln x_i \tag{6.1}$$

This equation, which does not contain temperature, is valid for the formation of one mole of mixture of any number of components, i, in whatever single state of aggregation, provided that the molecules of all the components retain the same indifference, or equivalence, to each other as the molecules of ideal gases. This is of course, seldom so, but equation (6.1) remains the expression of the fundamental contribution to the entropy of mixing, the statistical aspect of which has been touched upon in sections 4.7.2 and 4.7.3. Since x_i is a fraction, $\ln x_i$ is negative, so, of course, ΔS_{mix} is positive. This is why, it will be remembered, 'pure' is an essential word in the statement of the third law.

It is useful to inspect a graphical representation of equation (6.1), such as given in Fig. 6.1, for a binary mixture. It is seen that as x_i becomes very small, its proportional effect increases. The limiting tangents are vertical. This is why hardly anything is *quite* insoluble in anything else.

6.2.2   Partial molar extensive properties

Formally to deal with mixtures and solutions, closed systems are discarded in favour of open systems, i.e. systems no longer constant in material content. Additional independent variables are required. Thus,

$$X = X(T, P, n_1, n_2, n_3 \ldots) \tag{6.2}$$

where X is any extensive property, and n_1, n_2, n_3 ... are *numbers of moles of components 1, 2, 3* ... in the system, as many such terms as necessary being included, but *no more*. If, for instance, substance 3 were produced solely by an equilibrium reaction between 1 and 2 ($1 + 2 \rightleftharpoons 3$), it would be a mistake to include n_3 in equation (6.2) for the system concerned. The essence of the definition of a component is its independence of all other components in the same system: n_1, n_2, n_3 ... in equation (6.2) must be *independent variables*.

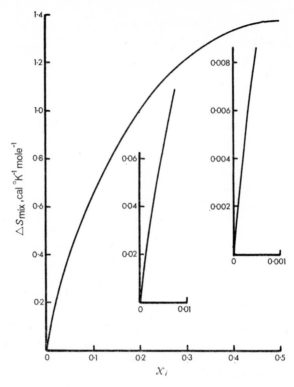

Fig. 6.1 Entropy of mixing, $\Delta S_{\text{mix}} = -R\Sigma x_i \ln x_i$, in cal $^{\circ}K^{-1}$ per mole of binary mixture.

X remains an extensive property, and dX remains a complete differential, so that one can write

$$dX = \left(\frac{\partial X}{\partial T}\right)_{P,n_1,n_2,...} dT + \left(\frac{\partial X}{\partial P}\right)_{T,n_1,n_2,...} dP +$$

$$+ \left(\frac{\partial X}{\partial n_1}\right)_{T,P,n_2,n_3,...} dn_1 + \left(\frac{\partial X}{\partial n_2}\right)_{T,P,n_1,n_3,...} dn_2 + ... \qquad (6.3)$$

all but the *one* appropriate independent variable being held constant in each partial differential coefficient. If temperature and pressure are kept constant,

$$dX = \left(\frac{\partial X}{\partial n_1}\right)_{T,P,n_2,n_3...} dn_1 + \left(\frac{\partial X}{\partial n_2}\right)_{T,P,n_1,n_3...} dn_2 + ... \qquad (6.4)$$

These partial differential coefficients are the partial *molar extensive properties of the components* concerned. Symbolisation is simplified by using

i for the component under consideration, and j for all the others. $\bar{X}_i$ is used to denote the partial differential coefficient, i.e. the 'partial molar X of component i'. Thus,

$$\bar{X}_i = \left(\frac{\partial X}{\partial n_i}\right)_{T,P,n_j} \tag{6.5}$$

X may be E, H, V, S, A or G. For a pure substance, partial molar X is the same as molar X; $\bar{X}_i = X_i^\circ$. $\bar{X}_i$ has the dimensions of X per mole. Fixed by nature for a pure substance at constant T, P, it becomes a function of composition when this same substance is mixed with another.
Equation (6.4) can be written

$$dX = \bar{X}_1\, dn_1 + \bar{X}_2\, dn_2 + \dots \tag{6.6}$$

Integration, with T, P and composition retained constant, gives

$$X = \bar{X}_1 n_1 + \bar{X}_2 n_2 + \dots \tag{6.7}$$

which is a statement of the obvious. General differentiation gives

$$dX = \bar{X}_1\, dn_1 + n_1\, d\bar{X}_1 + \bar{X}_2\, dn_2 + n_2\, d\bar{X}_2 + \dots \tag{6.8}$$

so that comparison with equation (6.6) indicates that

$$n_1\, d\bar{X}_1 + n_2\, d\bar{X}_2 + \dots = 0 \tag{6.9}$$

which is the *Gibbs–Duhem equation*. This equation expresses the fact that the partial molar properties of the components of a mixture are not susceptible to arbitrary variation, independently of each other. It may be regarded as a criterion of equilibrium, and is not subject to infringement.

Expressed more generally for *one mole* of mixture (i.e. a system containing Avogadro's number of molecules of whatever kind), equations (6.6), (6.7) and (6.9) assume the succinct forms

$$dX = \Sigma \bar{X}_i\, dn_i \tag{6.10}$$

$$X = \Sigma n_i\, \bar{X}_i \tag{6.11}$$

$$\Sigma n_i\, d\bar{X}_i = 0 \tag{6.12}$$

If X is entropy.

$$\Delta S_{\mathrm{mix}} = \Sigma x_i(\bar{S}_i - S_i^\circ) \tag{6.13}$$

and, ideally,

$$= -R\Sigma x_i \ln x_i \tag{6.1}$$

which is positive, so that in this case $\bar{S}_i > S_i^\circ$, as to be expected. Superimposed, there may be contributions of either sign due to the effects of specific interactions existing between like molecules before mixing, and between unlike molecules afterwards. The requirement for spontaneous mixing is of course that

$$\Delta G_{\text{mix}} = \Delta H_{\text{mix}} - T\Delta S_{\text{mix}} \tag{6.14}$$

shall be negative.

All the $\bar{X}_i$ quantities are related to each other in the same ways as the X quantities—there are no new equations to learn. Thus, $\bar{H}_i = \bar{E}_i + P\bar{V}_i$, $\bar{A}_i = \bar{E}_i - T\bar{S}_i$, $\bar{G}_i = \bar{E}_i + P\bar{V}_i - T\bar{S}_i$. It follows that in a mixture, $\bar{A}_i < A_i^\circ$ and $\bar{G}_i < G_i^\circ$. On the other hand, it is possible for $\bar{E}_i = E_i^\circ$, $\bar{H}_i = H_i^\circ$ or $\bar{V}_i = V_i^\circ$. In words, mixtures can be formed without total volume change and without evolution or absorption of heat, but never without entropy change and, in consequence, never without free energy change.

We thus have the remarkable circumstance that while there is no agreed basis for defining an 'ideal liquid', there is a sound basis (indeed, more than one, as will be seen) for defining an ideal liquid mixture. It rests on $\Delta H_{\text{mix}} = 0$, $\Delta V_{\text{mix}} = 0$, $\Delta S_{\text{mix}} = -R\Sigma x_i \ln x_i$, and, in consequence of equation (6.14), $\Delta G_{\text{mix}} = RT\Sigma x_i \ln x_i$.

There are sub-classifications of non-ideal liquid mixtures (which represent the vast majority). Those for which ΔS_{mix} is ideal are called 'regular'; those for which $\Delta H_{\text{mix}} = 0$ are called 'athermal', but this field cannot be entered beyond reference to an authoritative text.[1]

6.2.3   Partial molar volumes

This subject provides the first of some correlated interludes to the formal theme.

Partial molar volumes are derived from *densities* by methods (not discussed) subject to the disability of sensitivity to errors (cf. section 5.5.3). Results of only moderate accuracy may require large numbers of careful measurements. Ostensibly, a dull and unrewarding field. The reasons why it has been transformed quite recently into one of enthusiastic activity are interesting. There has been a rapid growth of concern with the structure of water,[2] and with the interactions between water on the one hand and, on the other, surfaces and solutes of all kinds. These are of primary importance in the life sciences. The problems concerned are complex, and must be approached by the study of comparatively simple 'model systems', such as aqueous solutions of organic compounds—even hydrocarbons. Other developments have contributed; notably computers have made new methods of calculation practicable,[3] and this in turn has given point to the

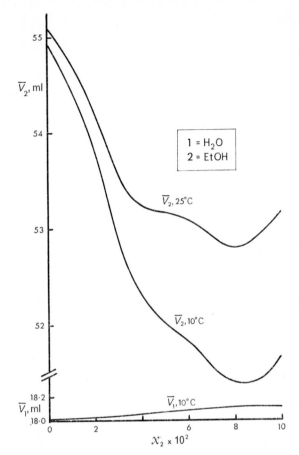

Fig. 6.2   Partial molar volumes, ml mole^{-1}, of the components in aqueous ethanol
solutions.

development of new experimental methods of vastly improved accuracy—in
the case of densities to a level of precision around 1 ppm.

Fig. 6.2 gives an inkling of the interest latent in partial molar volume data
on this new plane. It relates to solutions of ethanol in water up to a mole
fraction of 0·1; the contraction on mixing these two components has long
been known (and deplored), but possible explanations are recent.[4] This
particular problem will be taken up again in following 'interludes', but
occasion is taken to point out that $\overline{V}_1$ and $\overline{V}_2$ at the same temperature conform to the Gibbs–Duhem equation, as intelligent (and not too critical)
inspection of the diagram will show.

6.2.4   Chemical potential of a component

The quotation of but two independent variables is not adequate to define the state of an open system; as many additional independent variables are required as the number of components the system contains. It is also apparent that the expression of the first law valid for a closed system, namely

$$dE = TdS - PdV$$

will not do for an open system. It must be expanded to include the effects of infinitesimal variation in content of components, i.e.

$$dE = TdS - PdV + \Sigma\mu_i \, dn_i \qquad (6.15)$$

where dn_i represents an infinitesimal addition of dn_i mole of component i. This is an increment in a capacity factor; μ_i is the appropriate intensity factor for component i, analogous to the other intensity factors, T and P. Products of this nature, with the over-all dimensions of energy, must be included for each component. Equation (6.15) is, then, the first of the four fundamental equations of Gibbs appropriate to open systems (cf. section 5.4). Development of the other equations, as previously, leads to

$$dH = TdS + VdP + \Sigma\mu_i \, dn_i \qquad (6.16)$$

$$dA = - SdT - PdV + \Sigma\mu_i \, dn_i \qquad (6.17)$$

$$dG = - SdT + VdP + \Sigma\mu_i \, dn_i \qquad (6.18)$$

Equations (6.15) to (6.18) lead to alternative expressions for μ_i, the chemical potential of component i, but attention need be given only to the last of them, namely,

$$\mu_i = \left(\frac{\partial G}{\partial n_i}\right)_{T,P,n_j} \qquad (6.19)$$

which is seen to be identical with the partial molar Gibbs free energy, $\bar{G}_i$.

The next step is to show that the word 'potential' applied to this function is justified. The appropriate form of equation (6.10) is

$$(dG)_{T,P} = \Sigma\mu_i \, dn_i \qquad (6.20)$$

which lends itself to the previously used test for equilibrium. Consider the infinitesimal transfer of dn_i mole of component i from situation A, where its chemical potential is μ_i^A, to situation B, where its chemical potential is μ_i^B, the transfer being conducted at constant T and P. Then,

$$(dG)_{T,P} = \mu_i^B \, dn_i - \mu_i^A \, dn_i$$

$$= (\mu_i^B - \mu_i^A) \, dn_i$$

If $\mu_i^B = \mu_i^A$, $(dG)_{T,P} = 0$, demonstrating equilibrium. If $\mu_i^B < \mu_i^A$, $(dG)_{T,P} < 0$, and transfer of i from A to B will be a spontaneous process, and will tend to occur until $\mu_i^B = \mu_i^A$.

Excluding the intervention of other potentials, yet to be met, this remains true whether the two 'situations' differ physically or chemically. The transfer could be from one region to another within the same phase, or from one phase to another (as from liquid to vapour), or from one state of chemical combination to another (as of oxygen from the gaseous state to the combined state in an oxide). The transfer can thus be transport by diffusion, or it may be evaporation, condensation, crystallisation, solution or chemical reaction. In all cases component i tends to pass from higher to lower μ_i. For equilibrium within a system, μ_i for component i must be uniform throughout the system. The designation 'potential' is therefore accurate—but it must be remembered that μ_i is a property of a specific component i, and is expressed in cal mole^{-1}. It can equally well be used in relation to closed as to open systems. For a pure component, μ_i is identical with G_i, and μ_i° with G_i°. The most general, and preferred, expression for ΔG of chemical reaction is

$$\Delta G = \Sigma n_i \mu_i \,(\text{products}) - \Sigma n_i \mu_i \,(\text{reactants}) \qquad (6.21)$$

The intention is next to illustrate, without elaboration, the use of μ_i in a number of connections.

6.3   Applications of μ_i

6.3.1   The Gibbs phase rule

The rule has two aspects. First, it lays down the minimum number of statements needed to fix the equilibrium state of a system made up of C components shared between P phases. This number is called the *variance*, symbolised F. Secondly, it restricts the number of phases that can coexist in equilibrium, and defines the number of conditions that can be independently varied without a change in the number of phases in equilibrium. This is called the *number of degrees of freedom*—a little thought confirms that it is the same as the variance.

Deriving the rule is a matter of counting the number of independent variables sufficient uniquely to determine the state of the system, carefully excluding variables that, *by reason of internal equilibria known to be established*, the system determines for itself. Temperature and pressure account for two independent variables. The others to be considered relate to composition.

Taking one of the P homogeneous phases, it would seem to be necessary to specify $C - 1$ mole fractions, x_i, to define its composition, bearing in mind that $\Sigma x_i = 1$. For all P phases, with temperature and pressure, this would lead to a total of $2 + P(C - 1)$ statements to be made. This number is, however, greater than necessary. There is a unique, equilibrium distribution of each component over all the phases, so that the phases do not have independent compositions. Denoting the phases α, β, γ, δ ..., the fact of equilibrium requires that for component i

$$\mu_i^\alpha = \mu_i^\beta = \mu_i^\gamma = \mu_i^\delta \ldots$$

i.e. the chemical potential of component i must be everywhere the same. This must be true, severally, for all the components. There are, therefore, $P - 1$ identities of this kind for each component. Since in each phase μ_i is an explicit, continuous function of x_i, $C(P - 1)$ of the mole fractions originally listed are not independent, and must be excluded from the count of independent variables. Hence

$$F = 2 + P(C - 1) - C(P - 1)$$

Therefore

$$F = C - P + 2 \tag{6.22}$$

which is the usual expression of the phase rule. It is a simplifying and guiding principle, of inestimable value in chemistry, metallurgy and materials science generally. The present context does not allow development of the theme, beyond pointing to the crucial importance of definitions in statement and application of the rule. The *independence* of a component has already been sufficiently emphasised, but perhaps the *homogeneous* nature of a phase needs comment. A phase may, of course, be a mixture, but, if it is, the mixing must be of the full molecular randomness permitted by the nature of the components—entropy of mixing must be maximal.

6.3.2   Phase equilibria in one-component systems

Three phases of a pure substance coexisting in equilibrium constitute an invariant system $(F = 1 - 3 + 2 = 0)$ realisable only at a unique *triple point*; neither temperature nor pressure can be varied without disappearance of one of the phases. Two phases of a pure substance in equilibrium form a univariant system $(F = 1 - 2 + 2 = 1)$; at each temperature there is an equilibrium pressure determined by the system itself, or vice versa. For such a system, equilibrium between the two phases (e.g. liquid α, saturated vapour β) requires identity of chemical potential, represented by $\mu_i^\alpha = \mu_i^\beta$. Suppose that an infinitesimal change of temperature be imposed; the system must respond with an appropriate infinitesimal change of pressure to restore equilibrium and re-establish equality of μ_i between the phases. When this is accomplished, the infinitesimal changes of μ_i must also be the same, i.e. $d\mu_i^\alpha = d\mu_i^\beta$. Expressing these increments as functions of temperature and pressure,

$$d\mu_i^\alpha = \left(\frac{\partial \mu_i^\alpha}{\partial T}\right)_P dT + \left(\frac{\partial \mu_i^\alpha}{\partial P}\right)_T dP = d\mu_i^\beta = \left(\frac{\partial \mu_i^\beta}{\partial T}\right)_P dT + \left(\frac{\partial \mu_i^\beta}{\partial P}\right)_T dP$$

Bearing in mind that dG is a complete differential, so that, for example,

$$\frac{\partial^2 G}{\partial n_i\,\partial P} = \frac{\partial^2 G}{\partial P\,\partial n_i}$$

we can identify the partial differential coefficients and rewrite the equality as

$$-\bar{S}_i^\alpha\, dT + \bar{V}_i^\alpha\, dP = -\bar{S}_i^\beta\, dT + \bar{V}_i^\beta\, dP$$

Since for a pure substance the partial molar and the molar quantities are the same, rearrangement gives

$$dP(V^\alpha - V^\beta) = (S^\alpha - S^\beta)\, dT$$

or
$$\frac{dP}{dT} = \frac{\Delta S}{\Delta V}$$

recognised as the Clapeyron–Clausius equation, alternatively derived.

6.3.3   Ideal binary mixtures of volatile liquids

For isothermal mixing at constant pressure, we can write (using a more convenient notation than previously)

$$\Delta G^M = \Delta H^M - T\Delta S^M \tag{6.23}$$

If the mixing is ideal, $\Delta H^M = 0$, so that $\Delta G^M = -T\Delta S^M$. Then, for the formation of one mole of ideal binary mixture, using equation (6.1),

$$\Delta G^M = RTx_1 \ln x_1 + RTx_2 \ln x_2 \tag{6.24}$$

But it is evident that

$$\Delta G^M = x_1(\mu_1 - \mu_1^\circ) + x_2(\mu_2 - \mu_2^\circ) \tag{6.25}$$

where, for example, μ_1 is the chemical potential of component 1 in the mixture, and $\mu_1^\circ(= G_1^\circ)$ is its chemical potential in the pure state before mixing. It follows, by comparison of equations (6.24) and (6.25), that

$$\mu_1 - \mu_1^\circ = RT \ln x_1 \quad \text{and} \quad \mu_2 - \mu_2^\circ = RT \ln x_2$$

In general, for an ideal mixture,

$$\mu_i = \mu_i^\circ + RT \ln x_i \qquad (6.26)$$

an equation which is indeed an alternative definition of an ideal mixture.*

If the binary mixture is of two volatile liquids, it comes into equilibrium with a mixed vapour phase, and each component has the same chemical potential in both liquid and vapour phases. This is true for all compositions of the mixture, and for the pure components. Treating the vapour as a mixture of ideal gases, then, in the vapour, $\mu_i = G_i$ is given by

$$\mu_i = \mu_i^\circ + RT \ln (P_i/P_i^\circ) \qquad (6.27)$$

where P_i is the partial pressure of component i in the vapour, and P_i° is a standard pressure. Choosing this to be the vapour pressure of the pure component i, and comparing with equation (6.26), μ_i is the same in each, μ_i° is the same in each, so that

$$x_i = P_i/P_i^\circ \, ,$$

or $$P_i = x_i P_i^\circ \qquad (6.28)$$

which is *Raoult's law*. This is the classical definition of an ideal liquid mixture or solution; the equilibrium partial pressure of a component is proportional to its mole fraction, the proportionality constant being the vapour pressure of the pure component at the same temperature. If one component in a binary mixture obeys Raoult's law over the whole composition range, the Gibbs–Duhem relation requires the other to do so as well.

6.3.4   The ideal dilute solution

There is no fundamental distinction between solvent and solute. They are invariably called components 1 and 2 respectively, and $x_2 \ll x_1$. The ideal dilute solution is strictly attained only by infinite dilution of a real solution.†
Relationships derived for the one involve approximations—always to be looked for—when applied to the other.

The main topic is that of 'colligative properties'. These are properties depending solely on x_2, the mole fraction of the solute, irrespective of its nature, provided that it does not dissociate or associate in solution. The essential requirement is to be able to define a mole of solute—Avogadro's

---

* There is no occasion to worry about what happens as x_i tends to zero; x_i and $\ln x_i$ accompany each other (as in equation (6.24)), and x_i tends to zero faster than $\ln x_i$ tends to $-\infty$.

† It has been forcibly pointed out to the writer that 'infinite dilution' and 'zero concentration' as limiting cases carry different implications. This is worth thinking about.

number of kinetically independent solute particles—in the dissolved state. Since for a solution of a single substance $x_1 + x_2 = 1$, these properties are equally functions of x_1, and, since ideality is postulated, of $\mu_1 - \mu_1^\circ$, in terms of the equation

$$\mu_1 = \mu_1^\circ + RT \ln x_1$$

where μ_1 and μ_1° are the chemical potentials of the 'solvent substance' in the solution and in the pure solvent, respectively, at the temperature T.

This provides a route to the study of substances in solution, which, classically and on first encounter by students, resolves itself into the dull subject of the determination of the molecular weights of substances in solution. There is one general principle. It is to set up an experimental system that allows establishment of 'solvent substance equilibrium' between the solution and the pure solvent in a separate phase. This involves, necessarily, using μ_1 (pure solvent) $= \mu_1(T, P)$, and when accomplished, gives $\mu_1 - \mu_1^\circ$, and, in turn, x_1 and x_2. The latter, if the weight fraction of the solute in the solution is known, leads to the required molecular weight.

The common methods are classifiable according to the state of aggregation of the solvent in the separate pure phase, as in Table 6.1.

Table 6.1 Methods of determining molecular weights of substances in solution

Separate pure solvent phase	Method
Vapour	Relative lowering of vapour pressure
Vapour	Elevation of boiling point
Solid	Depression of freezing point
Liquid	Osmotic pressure

The first of these methods rests on application of Raoult's law (see equations (6.27) and (6.28)) and needs no explanation beyond

$$(P_1^\circ - P_1)/P_1^\circ = 1 - P_1/P_1^\circ = 1 - x_1 = x_2$$

The second and third methods are best considered together in terms of μ_1, formally represented as a function of temperature in Fig. 6.3.

Threelines, in tersecting at freezing—and boiling—points (T_f° and T_b°, where μ_1 is single-valued for the pair of phases in equilibrium), have slopes, $(\partial \mu_1/\partial T)_P = -\bar{S}_1$, becoming more negative with solid (α) $\rightarrow$ liquid (β) $\rightarrow$ vapour (γ) transitions, as expected from the smaller and larger entropy jumps attending melting and boiling, respectively.* If now a solute be added to the

---

* It may be disconcerting that μ_1, said to quantify 'escaping tendency' decreases with rising temperature. But it is to be noted that the rate of decrease increases in the sequence solid, liquid, gas. The escaping tendency of the component *from* the more confined state *to* the less confined state therefore increases with rising temperature.

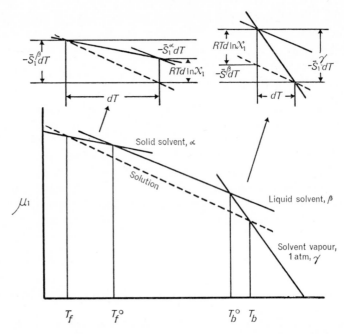

Fig. 6.3   Formal representation of the origins of freezing-point depression and boiling-point elevation.

liquid solvent phase, the chemical potential of the solvent is changed by $\mu_1 - \mu_1^\circ = RT \ln x_1$, and becomes as represented by the broken line in Fig. 6.3. This cuts the lines for pure solid and pure vapour solvent phases at the new freezing and boiling points, T_f and T_b, indicating how the two-phase systems (solution, pure solid solvent; solution, pure solvent vapour at 1 atm) respond to the solute addition, re-establishing 'solvent substance equilibrium' by spontaneous change of temperature. It is evident why, for a given solution, freezing-point depression is greater than boiling-point elevation.

Inspection of Fig. 6.3 shows that, at the limit of dilution,

$$RT \, d(\ln x_1) - \bar{S}_1^\alpha \, dT = -\bar{S}_1^\beta \, dT$$

or $$RT \, d(\ln x_1) - \bar{S}_1^\beta \, dT = -\bar{S}_1^\gamma \, dT$$

Then, for freezing,

$$- RT \, d(\ln x_1) = (\bar{S}_1^\beta - \bar{S}_1^\alpha) \, dT = \Delta S_f \, dT = (\lambda_f / T_f) \, dT$$

where λ_f is molar latent heat of fusion and dT is freezing-point *depression*, counted positive. Since the case for boiling at 1 atm pressure is similar, we can generalise:

$$- d(\ln x_1) = -(1/x_1)dx_1 = (\lambda/RT^2)\, dT$$

where dT is either freezing-point depression, or boiling-point elevation.

Since in the limit, $x_1 \to 1$, and since $x_1 + x_2 = 1$, so that $dx_1 + dx_2 = 0$,

$$dx_2 = (\lambda/RT^2)\, dT$$

Since, further, $x_2 = n_2/(n_1 + n_2) \to n_2/n_1 = m_2 M_1/1000$, where m_2 is molality of solute and M_1 is molecular weight of solvent,

$$dx_2 = (M_1/1000)\, dm_2, \text{ so that}$$

$$\frac{dT}{dm_2} = \frac{RT^2}{\lambda} \cdot \frac{M_1}{1000} = \frac{RT^2}{1000l} = K_f \text{ or } K_b \qquad (6.29)$$

where K_f or K_b are cryoscopic or ebullioscopic constants, T and l being appropriate temperatures and latent heats *per gramme* of solvent for fusion, or evaporation at 1 atm pressure. Despite the approximations entering on departure from infinite dilution, T is linear with m_2 over a range tolerable for the limited purpose of molecular weight determination. Quite different methods are necessary in applying freezing-point depression to the study of the non-ideality of solute behaviour—a subject yet to be discussed.

The osmotic pressure method depends on a different means of self-adjustment of the chemical potential of solvent substance. A system consisting of pure, liquid solvent, separated from a solution by a semipermeable membrane—permeable to solvent but not to solute molecules—is not, at uniform temperature and pressure, in a state of equilibrium. There will be a natural, osmotic flow of solvent from the region where its chemical potential is higher to the region where its chemical potential is lower, i.e. from the pure solvent (μ_1°) across the membrane, into the solution ($\mu_1 = \mu_1^\circ + RT\ln x_1$). If the compartment containing the solution is closed, this flow will *generate* an excess pressure on the solution side of the membrane—and also dilute the solution. To study the effect quantitatively, therefore, the excess pressure is found which, applied externally, brings osmotic flow to zero. This is called the osmotic pressure of the solution, π. The system, with pure solvent at 1 atm, and solution at $\pi + 1$ atm pressure is then in equilibrium, the chemical potential of solvent substance being the same, μ_1°, on either side of the membrane. That the decrease in μ_1 due to addition of solute has been nullified by the increase of pressure can be represented for the limiting case of the infinitely dilute solution by

$$RTd(\ln x_1) + \left(\frac{\partial \mu_1}{\partial P}\right)_T dP = 0$$

Therefore

$$-d(\ln x_1) = - (1/x_1)\, dx_1 \rightarrow -dx_1 = dx_2 = (\overline{V}_1/RT)\, dP$$

Therefore

$$dP = (RT/\overline{V}_1)\, dx_2$$

Treating $\overline{V}_1$ as constant, independent of pressure and x_2, integration between limits of $P = 1$ and π, and $x_2 = 0$ and x_2, gives

$$\pi = (RT/\overline{V}_1)x_2$$

Using $x_2 \rightarrow n_2/n_1$ in the limit of dilution,

$$\pi n_1 \overline{V}_1 = n_2 RT$$

Since $n_1 \overline{V}_1 \rightarrow V$, the volume of solution containing n_2 mole of solute,

$$\pi V = n_2 RT \tag{6.30}$$

which is the van't Hoff equation of 1885, identical in form with the general ideal gas law. It was this coincidence that led to the dissemination in elementary teaching of the nonsense that osmotic pressure is due to 'solute molecule bombardment', fortunately now dying out. It is so important, particularly in biology, to understand that isotonic solutions (i.e. of equal osmotic pressures) are those of equal chemical potential of the solvent. Perhaps it is the main message of this subsection that colligative properties of solutions are fundamentally, and most expeditiously, to be considered in terms of the chemical potential of the solvent.

6.4   Non-ideal liquid mixtures and solutions

6.4.1   Binary mixtures of volatile liquids

Conformity with Raoult's law is confined to mixtures of closely similar components (e.g. ethylene and propylene dibromides), and is not a sufficient criterion of ideality. Stricter criteria—zero heats and volumes of mixing—are still more rarely satisfied. Non-ideality, then, is the general rule, with deviations from Raoult's law either negative ($P_i < x_i P_i^{\circ}$) or, much more frequently, positive ($P_i > x_i P_i^{\circ}$); two examples are shown in Fig. 6.4. It is natural to ask for reasons for the deviations, and, in particular cases, plausible answers are at hand. For instance, hydrogen-bonding between the (activated) hydrogen atom of chloroform and the carbonyl oxygen atom of acetone (Fig. 6.4a) is to be expected. The second example illustrated

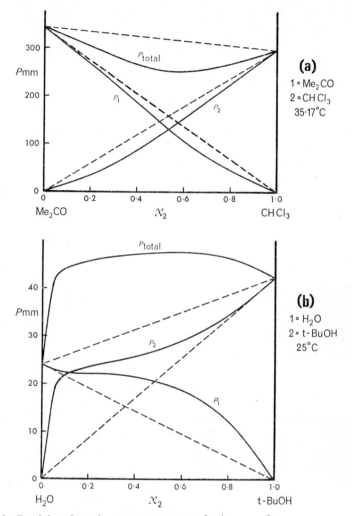

Fig. 6.4 Partial and total vapour pressures of mixtures of
(a) acetone and chloroform at 35·17 °C.
(b) water and t-butyl alcohol at 25 °C.

(Fig. 6.4b) presents a tougher problem, since both components are themselves hydrogen-bonded liquids of by no means simple structure.* In such
cases, clues may be obtained by comparing systems. Thus, methanol (1) and

* An experiment is recommended for those who find a conceptual difficulty about
'structure' in liquids; it is to mix acetonitrile and water in a boiling tube. There is a strong
cooling effect. How is such highly endothermal mixing to be explained?

carbon tetrachloride (2) behave in similar fashion, giving a diagram very like Fig. 6.4b, but rather more symmetrical. Methanol is an 'associated', hydrogen-bonded liquid. Carbon tetrachloride is not—it is 'non-associated' and non-polar, and, added to methanol, acts as a diluent. It separates methanol molecules from each other, breaks their hydrogen-bonding, and consequently increases their escaping tendency. Conversely, the strong attraction between methanol molecules opposes this invasion by carbon tetrachloride molecules, and increases *their* escaping tendency by a 'squeezing out' effect. Very likely, then, tertiary butyl alcohol acts similarly as a diluent to water—at least, this may be one factor contributing to the behaviour of this more complex system. But only limited progress can be based on such conjectural discussion, and a broader view must be taken to understand the nature of the problems to be faced.

Liquids have already been presented (section 4.3) as energy-entropy battlegrounds, where fluctuations have full play. The simplest model of a normal, unassociated liquid,[2,5] not much above its freezing point, is that of an *irregularly* close-packed system of molecules in a state of constant, 'flickering' rearrangement. The average co-ordination (i.e. number of nearest neighbours per molecule) is 10 to 11, but decreases with rise of temperature. Water, with an average co-ordination of about 4·5, *increasing* with rise of temperature, is highly anomalous.[2] In general, the short-range intermolecular attractions, varying in kind and intensity from one liquid to another, are very sensitive to intermolecular distance. The inference can be made that the net effect of the attractive forces, conferring cohesion, will be sensitive to *volume*, and in turn, to temperature and pressure. This is so, even if not quite for the most apparently direct reason. It will be recalled from section 4.3 (along with the eccentricities of $C_P - C_V$) that the volume of a liquid typically increases threefold between freezing and critical temperatures. Yet the nearest neighbour distance remains almost constant, so it is the decrease in co-ordination and the proliferation of '*holes*' that are the main effects of raising temperature. These holes, or 'fluidised vacancies', are the key features of Eyring's theory of liquids,[6] so convincingly successful, especially in relation to the processes of self-diffusion and fluid flow, which proceed by activated jumps.

The briefest review of the nature of liquids shows that conclusions should not too readily be drawn as to what happens when two liquids are mixed. Even if the mixing is isothermal, the components may initially be at different reduced temperatures (T/T_C, where T_C is critical temperature), and therefore not in corresponding states. If mixing is athermal ($\Delta H^M = 0$), there may be wide negative deviations from Raoult's law, due to an excess entropy of mixing of otherwise similar molecules differing greatly in size. Even the mixing of non-polar molecules of similar size may involve such large positive deviations as to give rise to a miscibility gap (fluorocarbons and hydrocarbons). In short, the situation is that much remains to be done and understood about liquid mixtures; they present a field of great difficulty, but major interest, in which there is much current activity[7] beyond the scope and level of present discussion.

In these circumstances of facing problems as yet elusive to detailed interpretation, it is an object lesson to find that all the systems must conform to basic thermodynamic rules, and that it is the evidence provided by thermodynamic data that must be examined first—these aspects are the only present concern.

It is informative first to look at the thermodynamic function that comprehends all the intermolecular forces—the internal pressure, $(\partial E/\partial V)_T$ (cf. section 4.3). It is measurable, amongst other methods, by use of equation (5.7), namely, $(\partial E/\partial V)_T = T(\partial P/\partial T)_V - P$; P, the external pressure, is comparatively negligible, $(\partial P/\partial T)_V$ is best measured directly but can be seen to be equal to α/β. For liquids at $25\,°C$ and 1 atm external pressure, internal pressures range from about 2000 atm (non-polar), through 3000–5000 atm (polar) to 20000 atm for water (despite its open structure). It is of interest that increasing external pressure (in the range to 10000 atm) normally causes internal pressure first to rise somewhat, then to fall, ultimately through zero to negative values—as repulsions become dominant over attractions. This reflects on a larger scale the law of force between individual molecules. But the main point is that these large and variable internal pressures reflect the strength and variability from one liquid to another of the short-range forces between *like* molecules. Leaving aside forces between unlike molecules and other significant factors, we should, intuitively, not expect ideal mixing for any pair of liquids without nearly identical internal pressures. It is certainly the case that dissimilarity in internal pressure militates against mutual solubility.[8]

Consider the molecules of each component in a mixture in relation to their environments in terms of the following scheme:

Pure component 1	Mixtures of 1 and 2	Pure component 2
1 (1)	1 (1, 2)	1 (2)
2 (1)	2 (1, 2)	2 (2)

where environmental molecules are indicated in brackets. The Raoult's law standard states are those of the pure components, represented by 1(1) and 2(2). It can be seen, without formal proof, that small addition of the 'foreign' component has little proportional effect on environment. Conversely, as x_2, for example, is indefinitely decreased, both P_1 and $dP_1/dx_1 = -dP_1/dx_2$

tend to P_1°. This is the first restrictive thermodynamic rule; Raoult's law, $P_1 = x_1 P_1^\circ$ is invariably valid as a limiting law for the component of mole fraction tending to unity. This is why the partial pressure curve of such a component (no matter how eccentric its behaviour elsewhere) must always run tangentially into the Raoult's law line towards the ordinate representing that pure component. This is seen to be so for the examples illustrated in Fig. 6.4, and is the necessary behaviour for the *solvent* in the ideal dilute solutions discussed in section 6.3.4.

If Raoult's law is to remain valid over the whole composition range for a given binary mixture, it can be seen that the requirement is that, for example, the molecules of component 1 must be indifferent to change in environment from (1) to (1,2) to (2). The *standard* chemical potential of component 1 (eminently a function of environment) must remain unchanged at the value appropriate to the pure component, so that, throughout, its chemical potential remains strictly determined by its mole fraction in accordance with

$$\mu_1 = \mu_1^\circ + RT \ln x_1 \tag{6.26}$$

Standard state = pure liquid state

It is not hard to see how exceptionally this can be satisfied.

Another glance at the above scheme, relating molecules and environments, shows the existence of limiting cases of another kind, namely, 2(1) and 1(2), corresponding to infinitely dilute solutions of one component in the other. In each case, a condition for *some sort* of ideality of behaviour is satisfied, namely, uniformity of environment, even if now the environment is foreign. A second restrictive thermodynamic rule is to be anticipated.

If, for clarity, attention is confined to one end of the composition range, and it is agreed that component 1 ($x_1 \to 1$) is to be called solvent, and component 2 ($x_2 \to 0$) solute, it is seen that the conditions 1(1) and 2(1) simultaneously apply to solvent and solute respectively. If, then, the solvent is approaching ideal behaviour in the sense of following Raoult's law, should we not expect the solute also to be tending to some kind of behaviour also to be described as 'ideal'? This is, of course, likely to be different (except by a fluke) from 'Raoult's law ideality' because the appropriate standard state— 2(1) instead of 2(2)—is *qualitatively* different.

This question is settled by reference to the Gibbs–Duhem equation— equation (6.12), which, for a two-component mixture, with $\bar{X}_i = \mu_i$ is

$$x_1 \, d\mu_1 + x_2 \, d\mu_2 = 0 \tag{6.31}$$

If both components are in equilibrium with a vapour phase of ideal gas properties, in which both components conform to $\mu_i = \mu_i^\circ + RT \ln P_i - RT \ln P_i^\circ$, where P_i° is *any* fixed standard pressure, $d\mu_i = RTd(\ln P_i)$.

Then equation (6.31) can be written

$$x_1\, d(\ln P_1) + x_2 d(\ln P_2) = 0 \tag{6.32}$$

But $x_1 + x_2 = 1$; $dx_1 + dx_2 = 0$, and this is the same as

$$x_1 d(\ln x_1) + x_2 d(\ln x_2) = 0 \tag{6.33}$$

Now suppose component 1 obeys Raoult's law as a limiting law, so that $P_1 = x_1 P_1^\circ$ where P_1° is the vapour pressure of the pure component 1, $\ln P_1 = \ln x_1 + \ln P_1^\circ$ so that $d(\ln P_1) = d(\ln x_1)$.
Equation (6.33) then becomes

$$x_1\, d(\ln P_1) + x_2\, d(\ln x_2) = 0$$

Comparison with equation (6.32) shows that

$$d(\ln P_2) = d(\ln x_2)$$

Integrating, $\ln P_2 = \ln x_2 + \ln k$, where $\ln k$ is an integration constant. Hence the limiting law for component 2 is

$$P_2 = kx_2 \tag{6.34}$$

where k is a constant. *If* component 1 follows Raoult's law over the whole composition range, so that P_1 is linear with x_1, then P_2 must also be linear with x_2 over the whole range. In this case the integration constant can be evaluated, since when $x_2 = 1$, $P_2 = P_2^\circ = k$. This means that if one component conforms to Raoult's law over the whole composition range, the other component must do so as well. In other cases, k in equation (6.34) is indeterminate; it is a constant to be found experimentally.

Equation (6.34) is known as Henry's law. It carries the implication that the partial vapour pressure curve of a component with mole fraction tending to zero must always run tangentially into a straight line—the 'Henry's law line' of slope k, greater or less than that of the Raoult's law line. As far as *limiting* laws are concerned, it could be said that Raoult's law is a special case of Henry's law. This second restrictive thermodynamic rule can be seen operating in Fig. 6.4, but is more formally discussed and illustrated in the next subsection.

The significance of Henry's law is illuminated by its history. Originally, in 1803, it related to the solubility of gases: the mass of gas dissolved by unit volume of liquid at constant temperature is proportional to the gas pressure. Alternatively, the Ostwald solubility coefficient is independent of pressure, or the ratio of concentrations of the dissolving species in gas and solution phases is constant. It became clear that this was but one aspect of a general law relating to the equilibrium distribution of a component between different phases—as of a solute between immiscible solvents (e.g. of iodine between water and carbon tetrachloride). It is that the ratio of concentrations at equilibrium, the *partition* or *distribution coefficient*, is constant, independently of either concentration. To examine the thermodynamic basis of this law, consider that there is at equilibrium identity of chemical potential of the distributed component in the two phases, α and β, i.e. $\mu_i^\alpha = \mu_i^\beta$. If in each phase this component behaves ideally in the sense that $\mu_i = \mu_i^\circ + RT \ln x_i$, then, at equilibrium,

$$\mu_i^{\circ\alpha} + RT \ln x_i^\alpha = \mu_i^{\circ\beta} + RT \ln x_i^\beta \qquad (6.35)$$

therefore

$$RT(\ln x_i^\alpha - \ln x_i^\beta) = RT \ln (x_i^\alpha/x_i^\beta) = \mu_i^{\circ\beta} - \mu_i^{\circ\alpha} \qquad (6.36)$$

Since the latter difference is one between two fixed standard state functions, it follows that, at constant temperature, x_i^α/x_i^β is constant. Thus, the partition coefficient depends for its value on a difference of *standard* chemical potentials of the distributed component due to its difference of *molecular environment* in the two phases. This is helpful in relation to the preceding discussion. It may also be noted that the 'partition law' is the basis of partition chromatography.

Needless to say, the law is apparently modified in practice because of the intervention of the non-ideality of real systems. Nevertheless, the law is fundamental in terms of chemical potential, and therefore finds application in the study of non-ideality, whether of physical or chemical origin. These are matters for study elsewhere.

We have therefore arrived at the conclusion that Henry's law *does* represent a kind of ideal behaviour of the *solute*, attainable, strictly, only in infinitely dilute solution. But there is still a worry about the standard states introduced with some slyness in equation (6.35). To rectify this sharp practice, we return to the proven limiting law, $d \ln P_2 = d \ln x_2$. Remembering that P_2 is a partial pressure of 'ideal gas vapour', so that $d\mu_2 = RTd(\ln P_2)$, we can write

$$d\mu_2 = RTd(\ln x_2)$$

and integrate it to get

$$\mu_2 = \text{const.} + RT \ln x_2$$

Setting x_2 equal to unity as a standard state, we can call the constant the standard chemical potential of component 2, and represent it by μ_2°, thus getting an equation identical in form with equation (6.26), which applies to the solvent. But on looking into this, recollecting that Henry's law requires the maintenance of the 2(1) environment, we find that μ_2° so defined requires simultaneously infinite dilution conditions for, and unit mole fraction of, component 2. This is sufficiently ridiculous to drive the student to the pub again, unless he has remembered the earlier statement that standard states are often hypothetical states, not physically realisable. There can be no valid objection to writing

$$\mu_2 = \mu_2^\circ + RT \ln x_2 \qquad (6.37)$$

Standard state = hypothetical, infinite dilution state of unit mole fraction, x_2

provided that it is understood to be a limiting law appropriate to $x_2 \to 0$; if it turns out to be applicable to solutions of $x_2 > 0$, this is an unearned bonus. In practice, mole fraction is not used to express composition for solutions. The standard state then becomes less apparently absurd, but remains physically unattainable. Equation (6.37) can be regarded as a generalised Henry's law, applicable in the limiting case to any solute, whether or not volatile.

Still the main restrictive thermodynamic rule has not been presented—only a special aspect of it. It is the Gibbs–Duhem equation, $\Sigma n_i d\overline{X}_i = 0$, which rules all systems, however eccentric, without approximations. For application to the partial vapour pressures of a binary liquid mixture, it assumes the form already noted, i.e.

$$x_1 d \ln P_1 + x_2 d \ln P_2 = 0 \qquad (6.32)$$

Division by $dx_1 = -dx_2$ gives

$$\left.\begin{array}{l} x_1 \, d(\ln P_1)/dx_1 - x_2 \, d(\ln P_2)/dx_2 = 0 \\[2mm] (x_1/P_1)(dP_1/dx_1) - (x_2/P_2)(dP_2/dx_2) = 0 \end{array}\right\} \qquad (6.38)$$

or

which are two forms of the *Duhem–Margules equation*—a differential equation seen to express a fundamental correlation between the *slopes* of the partial vapour pressure curves of the two components. Valid experimental results must satisfy this requirement, in the manner shown in Fig. 6.5a. In this diagram, tangents are drawn at points x_1, P_1 and x_2, P_2, for one given mixture. Their slopes are dP_1/dx_1 and dP_2/dx_2. Chords are also drawn, joining the same points to their respective origins. The slopes of the chords are P_1/x_1 and P_2/x_2. The ratio of slopes, tangent/chord must be the same for each component, and this must be so for all mixtures. If

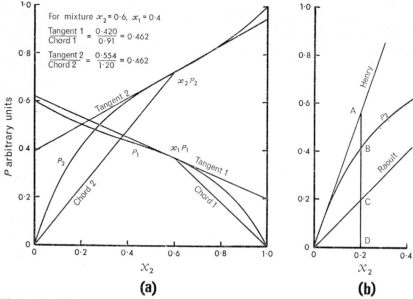

Fig. 6.5 (a) The Beatty and Callingaert test
(b) Raoult's law and Henry's law activity coefficients.

this test[9] fails, either the results are erroneous, or the assumption of vapour phase ideality has failed, and due correction must be applied.

Two of the implications of the Gibbs–Duhem equation in this field have now been noted. A third is that if one component shows positive deviations from Raoult's law over the whole composition range, the other component cannot show negative deviations over the whole composition range. This statement is framed to agree with a recent clarification[10] of some not uncommon misunderstandings. Further discussion requires use of functions formally adapted to quantify non-ideality.

6.4.2   Thermodynamic functions for non-ideality

The writer admits defeat in his intention to keep this section short, avoiding most of the tedious formalism. But the language must be learned, and the only choice is to attempt the job properly or to leave it alone. The reader's task is eased by further subdivision of a long section, starting with a statement on purposes.

6.4.2.1   *Ideality and non-ideality in perspective*

It is important to understand that the functions concerned are invented to isolate deviations from ideality in a form suitable for quantitative expression and subsequent enquiry into causes. They can be regarded as the thermodynamic raw material available for use in gaining closer understanding of physical systems, and in developing overly simple, idealised theoretical models into more realistic ones. Under extreme conditions, properties may become functions more of the deviations than of the ideal laws. It may then be expedient to set up a more appropriate standard of 'ideal' behaviour. Emphatically, 'ideal laws' and 'ideal behaviour'— together with 'standard states'—are matters of by no means unique *definition*. There are therefore numbers of functions expressing 'non-ideality'—all the more because of variations in conventions.

Introductory studies tend to be hag-ridden by the ideal gas and dilute solution; however unrepresentative, the simplest cases have to be considered first. To counter a possible impression that the next operation is to paper over cracks in not-so-ideal laws, it may be useful to put these matters into perspective. The chart in Fig. 6.6 is intended to help, together with the following *ad hoc* cogitation on gases.

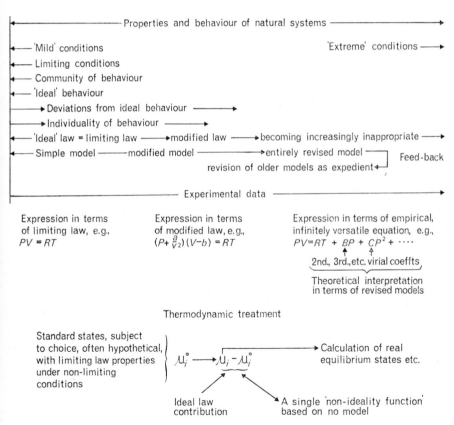

Fig. 6.6 'A scientific perspective'.

The choice of ideal law normally suited to gases is not questionable. The model is of randomly translating, dimensionless molecules interacting only in elastic collisions. Real gases conform at sufficiently low pressures. The van der Waals modified model deals fairly well with minor deviations at somewhat greater pressures. But for gases compressed to the densities of liquids, deviations become large and of major importance. A new theoretical approach is needed—in the broadest sense, a new model. A molecular 'cluster' theory, exacting in its mathematical development,[11] answers this description.

There is profit in more mature meditation on earlier studies. Reflection on Andrews isothermals, particularly in relation to the theorem of continuity of state, raises healthy questions. By what cycle of changes can a liquid be made to undergo successive ebullitions without intervening condensations? What is the difference between a liquid and a highly compressed gas? What is the significance of the critical temperature, above which a substance may exist as a solid, but not as a liquid? Does the latent heat fall precisely to zero at T_c, or is it replaced at somewhat higher temperatures by a heat capacity anomaly, reflecting continued differentiation between 'liquid-like' and 'gas-like' states? Other questions will come into the enquiring mind, and there would be still more profit in searching for the answers, some known, some emerging. But the writer's only purpose at present is to combat the 'cut and dried' atmosphere (perhaps intrinsic to the 'lectures cum examinations' system) in which a great deal of physical chemistry is offered for students' absorption. In particular, the object is to debunk 'ideal' (with its meritorious connotation) and to dispel the denigratory implications of 'non-ideal', 'imperfect', 'deviation', 'correction', etc. The ideal system or law is that thought to be well understood—indeed cut and dried—and it is in the field of deviations and non-ideality that the interest lies, with the possibilities of advancing knowledge of the very wide range of conditions and behaviour of material systems of which we initially get so narrowed a view.

With this necessary emphasis on the fundamental status of the non-ideality functions, attention may be turned to thermodynamic treatments—not concerned with questions except to provide the basis for asking them. The general procedure, seen to give pride of place to μ_i, can be illustrated in terms of gases.

6.4.2.2 *Gases: the fugacity*

The equation

$$\mu_i = \mu_i^\circ + RT \ln P$$

applies to an ideal gas, but not, except as an approximation or a limiting case, to a real gas. G. N. Lewis suggested replacing P by a function called *fugacity*, f_i, such that the equation

$$\mu_i = \mu_i^\circ + RT \ln f_i \tag{6.39}$$

with the same μ_i° holds exactly. It is seen that

$$\mu_{i(real)} - \mu_{i(ideal)} = RT \ln (f_i/P) = RT \ln \gamma_i \tag{6.40}$$

where γ_i is called a fugacity coefficient, or activity coefficient. It incorporates the non-ideality, and tends to unity as P tends to zero. Lewis proposed fugacity as the proper measure of escaping tendency,[12] avoiding the embarrassment that μ_i tends to $-\infty$ at zero pressure.

Since, at constant temperature,

$$d\mu_{i(real)} = V_i dP = RT d \ln f_i$$

whereas

$$d\mu_{i(ideal)} = (RT/P)dP = RT d(\ln P)$$

$$d(\ln f_i) - d(\ln P) = \left(\frac{V_i}{RT} - \frac{1}{P} \right)dP$$

or $\qquad RT \ln \gamma_i = RT \ln (f_i/P) = \int_0^P \left(V_i - \frac{RT}{P} \right)dP \tag{6.41}$

These equations can be solved, given experimental 'PVT data', either by graphical integration, or by use of an equation of state fitted to the data. If, at not too high pressures ($\ngtr$ 50 atm for 'permanent gases'), $PV = RT + BP$ is adequate, the integral in equation (6.41) reduces to BP. It is to be noted that, in general, $f_i \neq V_i/RT \neq P_{ideal}$ (for N_2 at $0°$ C and $P = 1000$ atm, $f_i = 1839$ atm, $P_{ideal} = 483$ atm; for CO_2 at $60°$ C and $P = 300$ atm, $f_i = 112$ atm, $P_{ideal} = 508$ atm). For mixtures of gases, the situation is more complex, for the equation of state must contain virial coefficients, B_{11}, B_{22}, B_{12} dealing with contributions to non-ideality from interactions between like and unlike molecules. It is also to be noted that fugacity, like chemical potential, has no meaning unless attached to a specified component. Inspection of equation (6.39) suggests that these two functions are measures of the same quantity on alternative scales: $f_i = \exp\{(\mu_i - \mu_i^\circ)/RT\}$. It follows that, for a mixed system, 'total fugacity', unlike total pressure, is not an intelligible concept.

Equilibrium constants, K_P, in gaseous systems are strictly, and at high pressures, functionally, determined by fugacities and not by partial pressures.

It may occasion surprise that in preceding discussions saturated vapours have been freely treated as if they were ideal gases. Happily, in those fields, pressures are usually low enough for this approximation to be tolerable within experimental error, but sometimes it is not. The Duhem–Margules equation, as written in equation (6.38) then fails; rewritten in terms of fugacities it, does not.

6.4.2.3   *Activity coefficients, symmetrical and unsymmetrical*

Our main interest lies with liquid mixtures and solutions, rather than gases. The general procedure is the same in principle and is, indeed equally valid for gas systems. Ideal behaviour can be specified by

$$\mu_i = \mu_i^\circ + RT \ln x_i$$

whatever the choice of standard state (real; hypothetical) on which μ_i° depends. For non-ideal systems, not conforming to this equation except under limiting conditions, a quantity a_i, the *activity* of component i is defined, such that

$$\mu_i = \mu_i^\circ + RT \ln a_i \qquad (6.42)$$

holds exactly. Again,

$$\mu_{i(\text{real})} - \mu_{i(\text{ideal})} = RT \ln(a_i/x_i) = RT \ln \gamma_i \qquad (6.43)$$

where γ_i is an *activity coefficient*—a dimensionless quantity comprehending departure from ideality.

For binary systems studied over the whole composition range, the components are treated equivalently, the standard state for each being the pure component at the same constant temperature. This gives rise to a symmetrical system of activity coefficients. In each case, γ_i is the ratio of the real partial pressure of component i to that required by Raoult's law, i.e. $P_i/P_i^\circ x_i$. These 'Raoult's law activity coefficients' are represented in Fig. 6.5b by BD/CD.

For solutions, it is convenient to retain the same standard state for the solvent, but to use for the solute a standard state *qualitatively* that of infinite dilution, but nominally of unit mole fraction, as already indicated in relation to equation (6.37). This leads to an unsymmetrical system of activity coefficients—a Raoult's law activity coefficient, $\gamma_1 = P_1/P_1^\circ x_1$, for the solvent, used simultaneously with a Henry's law activity coefficient, $\gamma_2 = P_2/kx_2$ (represented by BD/AD in Fig. 6.5b) for the solute.

This seemingly perverse, lop-sided scheme is adapted to a fairly natural division between the two fields of interest—liquid mixtures and solutions. On the one hand, deviations from Raoult's law, and the Raoult's law activity coefficients are mainly functions of interactions *between* the two different components. In the language of solutions, we should say that they are essentially due to *solute–solvent interactions*. On the other hand, with concentration of solution increasing from the limit of near-zero, with solute environment yet virtually unchanged, the first deviations from Henry's law are usually attributable to interactions between solute molecules, i.e. to *solute–solute interactions*. This is certainly so for electrolytic solutes, and is the basis for theorisation about 'activity coefficients' (invariably Henry's law activity coefficients, unless otherwise stated) of solutes in dilute solutions, with solvent assumed to be unchanged in properties from the pure state.

6.4.2.4   *Solutions: composition scales, symbolisation*

The composition of a solution is not normally expressed as a mole fraction of solute, but in terms of molality, moles of solute per kg of solvent, or of molar concentration, moles of solute per litre of solution. Molality shares with mole fraction the advantage of independence of temperature, but has the disadvantage of being radically affected by change from one solvent to another. Molar concentration is dependent on temperature because of thermal expansion, but retains its rather more fundamental significance on change of solvent.

We have therefore to face considerable variation in usage with attendant danger of confusion. To avoid it, attention should be fixed on one solution (there, in a beaker, on the bench), with single-valued, unique chemical potential of solute, μ_i, determined by the laws of Nature. The difficulties, man-made, arise in expressing this μ_i as a function of the chemical potential of the same solute in an arbitrarily defined standard state, and of the composition of the solution. Thus, we have as alternatives

$$
\begin{aligned}
\mu_i &= \mu_i^\circ + RT \ln x_i + RT \ln \gamma_i \\
&= \mu_i^\circ + RT \ln m_i + RT \ln \gamma_i \\
&= \mu_i^\circ + RT \ln C_i + RT \ln \gamma_i
\end{aligned}
\qquad (6.44)
$$

μ_i is identically the same, in cal mole^{-1}, in each case, but x_i, m_i and C_i are different. The standard states, not physically realisable, are those with $x_i = 1$, $m_i = 1$ mole kg^{-1}, $C_i = 1$ mole l^{-1}, with, in each case $\gamma_i = 1$, i.e. the limiting value denoting (Henry's law) ideality attained only at infinite dilution. It is very important to realise that, for example, $a_i = \gamma_i m_i = 1$ mole kg^{-1} (as for 1·734 molal aqueous KCl solution) is not a standard state. The requirements is that m_i and γ_i are *separately* equal to unity.

For one and the same real solution, then, the three μ_i° and the three γ_i are different. All three modes of expression are in use for good reasons, and none can be discarded. We have to live with this. Transposition of data from one scale to another (Appendix 2) is cumbersome, and the densities needed for transfer to the C_i scale are often not available.

The question of symbols is a vexed one. The writer, having to make a decision, felt unable to adopt recent recommendations[13] (already criticised,)[14] advocating f_B, γ_B and y_B for activity coefficients 'of *substance* B' on x, m and C scales, respectively. This uses up too many symbols on three out of at least ten variants of effectively the same function. Proliferation of symbols adds to the list of arbitrary conventions to be kept in mind, and aids neither clarity of expression nor ease of reading. Only three out of twenty standard texts surveyed had used this scheme; by far the most widely used and recognised symbol for activity coefficient is γ, with, occasionally, a subscript x, m or C when necessary—which is seldom, because the activity coefficient is rarely far separated from the composition term to which it belongs. The writer's policy is therefore to keep to the simplest and most readily understood usage, and to employ γ_i for the activity coefficient of *component i*, with suitable qualification when required.

6.4.2.5   *Solutions of electrolytess*

Electrolytic solutes need special attention. If the stoichiometric concentration of an ionisable solute is C, dissociation of fractional degree α leads to a greater concentration of kinetically independent solute particles, charged or uncharged. To take a specific example,

$$Cr_2(SO_4)_3 \rightleftharpoons 2Cr^{3+} + 3SO_4^{2-}$$

$$(1 - \alpha)C \qquad 2\alpha C \qquad 3\alpha C \qquad \text{Total } (1 + 4\alpha)C$$

It is this enhanced concentration of solute entities, of whatever kind, that determines the colligative properties of the solution. The general notation is

$$A_{v_+}^{z+} B_{v_-}^{z-} \rightleftharpoons v_+ A^{z+} + v_- B^{z-}$$

$$(1 - \alpha)C \quad v_+\alpha C \qquad v_-\alpha C \qquad \text{Total } 1 + (v - 1)C$$

where $v = v_+ + v_-$, i.e. v(nu) is the total number of ions provided by complete dissociation of one 'molecule' (i.e. mole/N) of solute, made up of v_+ cations and v_- anions. For electroneutrality, $v_+z_+ = v_-z_-$. It can be seen that the classical van't Hoff factor, i ($\pi = i (n_2/V) RT$; cf. equation (6.30)) is $i = 1 + (v - 1)\alpha$, and for α increasing towards unity, it tends to v; e.g. 2 for NaCl, 4 for LaCl$_3$.

To consider the application of the thermodynamic treatment to such systems, it is advisable to go back to the definition of chemical potential of a component, i.e.

$$\mu_i = \left(\frac{\partial G}{\partial n_i}\right)_{T,P,n_j} \tag{6.19}$$

How can this be applied, for example, to an aqueous solution of sodium chloride? We know that this salt is completely ionised in solution, so that we have, in effect, a mixed solution, containing sodium ions and chloride ions, moving about as independent (if somewhat non-ideal) solute particles. Although this is so, the fact remains that the two kinds of ions, of opposite sign of charge, are not independent because the solution as a whole must be electrically neutral. Neither the sodium ions nor the chloride ions separately can be regarded as components because the *operation* represented by equation (6.19), with i representing one kind of ion, and j including the other, is impossible. This is the reason why the expression 'ionic component' should not be used—it can be replaced by 'ion constituent'.[15] Since there is no conceivable experimental method of getting round this impasse, the conclusion is that the chemical potential of a single ionic species is unknowable.

There is, of course, no objection to conducting two such operations as the above simultaneously in such a way that electroneutrality is continuously maintained. But this is just the operation involved in defining the chemical potential of the electrolyte as a whole.

An 'extrathermodynamic step' is taken in defiance of this difficulty. The determinable quantity is split, as convenient, into parts, which are assumed to be entirely normal except for unknowability. But the challenge must always be met of combining two or more such undeterminable quantities together to get something unambiguously measurable. For instance, all the properties of sodium chloride in solution in water are attributable to aqueous sodium and chloride ions. It is then reasonable to write

$$\mu_{NaCl} = \mu_{Na^+} + \mu_{Cl^-}$$

whether or not the contributions on the right are determinable. Since the ions are real enough to add their quotas to colligative properties, and also to be non-ideal, it is equally reasonable, having taken the first step, to treat them further like any other solutes, so that

$$\mu_{Na^+} = \mu_{Na^+}^\circ + RT \ln a_{Na^+}$$

$$\mu_{Cl^-} = \mu_{Cl^-}^\circ + RT \ln a_{Cl^-}$$

Then, $\mu_{NaCl} = \mu_{NaCl}^\circ + RT \ln a_{Na^+}a_{Cl^-}$

from which it is seen that the *product* of individually unknowable ion activities is determinable, and it is convenient to use a *mean ion activity*, which, using a simpler notation, is

$$a_\pm = \sqrt{a_+ a_-}$$

Activity coefficients, and a *mean ionic activity coefficient* may be defined

$$a_+ = \gamma_+ C_+ \; ; \; a_- = \gamma_- C_- \; ; \; \gamma_\pm = \sqrt{\gamma_+ \gamma_-}$$

and since $C_+ = C_- = C$

$$\mu_{\text{NaCl}} = \mu_{\text{NaCl}}^\circ + RT \ln a_{\text{NaCl}}$$

$$= \mu_{\text{NaCl}}^\circ + 2RT \ln a_\pm$$

$$= \mu_{\text{NaCl}}^\circ + 2RT \ln C + 2RT \ln \gamma_\pm$$

The whole of this argument could equally have been conducted with m in place of C, but it is understood that, for a given electrolyte, standard states, activities and activity coefficients would not be the same. For the generalised electrolyte $A_{\nu_+}^{z_+} B_{\nu_-}^{z_-}$ (equals 'salt', for short) the relationships are

$$\mu_{\text{salt}} = \nu_+ \mu_+ + \nu_- \mu_- \tag{6.45}$$

$$a_\pm = (a_+^{\nu_+} a_-^{\nu_-})^{1/\nu} \; ; \; \gamma_\pm = (\gamma_+^{\nu_+} \gamma_-^{\nu_-})^{1/\nu} \tag{6.46}$$

$$\mu_{\text{salt}} = \mu_{\text{salt}}^\circ + RT \ln a_{\text{salt}} \; ; \; a_{\text{salt}} = a_\pm^\nu$$

$$\mu_{\text{salt}} = \mu_{\text{salt}}^\circ + \nu RT \ln C + \nu RT \ln \nu_\pm \tag{6.47}$$

These relationships have been worked out for a strong electrolyte with $\alpha = 1$. What if the electrolyte is incompletely dissociated, $\alpha < 1$? Inspection of equation (6.45) shows that it already expresses the fact of equilibrium so that no change is necessary, except to recognise that the significance of $\gamma_\pm$ is altered. Thus, if, as in equation (6.47), the mean ion activity coefficient 'operates' on the total molar concentration C, it includes the non-ideality contribution attributable to incomplete dissociation. It is classed as a *stoichiometric activity coefficient*. If, on the other hand the degree of dissociation, α, has been independently determined (as perhaps by conductance measurements), so that the total concentration can be accurately split into $C_{\text{molecules}}$ and C_{ions}, activity coefficients for each part will be required. There are here some grounds for separate symbolisation, but this is up to the workers engaged in this field, who understand the task in hand. The reader may care to count up the number of activity coefficients that have so far emerged in this discussion, in relation to names and symbols. Apart from descriptive mouthfuls like 'mean molal ionic stoichiometric activity coefficient', there are no special names, except that, traditionally, mole fractional activity coefficients are called 'rational activity coefficients'.

6.4.2.6 Osmotic coefficients

Henry's law activity coefficients, γ_2, are satisfactory for description of solute misbehaviour, but γ_1, Raoult's law activity coefficients for the solvent, are rather ineffective. This is because solvent behaviour is always tangential to the ideal law as x_1 tends to unity—as it does for solutions within a composition range of interest. This means that γ_1 is close to unity, and is insensitive to the non-ideality of the major component. This can also be seen, and its importance appreciated, by writing the Gibbs–Duhem equation in alternative forms, i.e.

$$x_1 d\mu_1 + x_2 d\mu_2 = 0 \tag{6.31}$$

$$x_1 d(\ln a_1) + x_2 d(\ln a_2) = 0 \tag{6.48}$$

or, putting $a_1 = \gamma_1 x_1$, etc., and remembering $dx_1 + dx_2 = 0$

$$x_1 d \ln \gamma_1 + x_2 d \ln \gamma_2 = 0 \tag{6.49}$$

The two terms are of equal significance; thought about the implications for γ_1 and γ_2 as $x_1 \to 1$ and $x_2 \to 0$ indicates why γ_1 is not a very satisfactory 'non-ideality function' under these circumstances. This is particularly so when the Gibbs–Duhem relation is being used to study γ_2 in terms of experimental measurement of solvent activity, by methods shortly to be outlined.

The difficulty is reduced by using a non-ideality factor multiplying $\ln x_1$ instead of x_1; this is the 'rational osmotic coefficient', g, related to γ_1 as follows:

$$\mu_1 + \mu_1^\circ + RT \ln a_1 \begin{cases} \mu_1 = \mu_1^\circ + RT \ln \gamma_1 x_1 \\ \mu_1 = \mu_1^\circ + gRT \ln x_1 \end{cases} \tag{6.50}$$

$$\ln a_1 = \ln \gamma_1 x_1 = g \ln x_1 \; ; \; g = 1 - \frac{\ln \gamma_1}{\ln x_1} \; ;$$

$$\ln \gamma_1 = (g - 1) \ln x_1 \tag{6.51}$$

It is informative to see how the name 'osmotic coefficient' arose. It is recalled (section 6.3.4) that, for the 'ideal dilute solution', the differential effect of the presence of solute on solvent, $d\mu_1 = RT \, d(\ln x_1)$, is equalled, on equilibration with a pure solvent phase, by $d\mu_1 = -\bar{S}_1 \, dT$ (f.pt. depression, b.pt. elevation) or by $d\mu_1 = \bar{V}_1 \, dP$ (osmotic pressure). Discarding the hypothesis of ideality, it is seen that it is $d\mu_1 = RT \, d(\ln a_1)$, which is equalised by one or other of $-\bar{S}_1 dT$ or $\bar{V}_1 dP$. Then, without approximation, for example,

$$gRT d(\ln x_1) + \bar{V}_1 dP = 0$$

or $$d(\ln x_1) = -\frac{\bar{V}_1}{gRT} dP$$

Integration between appropriate limits and rearrangement gives

$$\pi = -\frac{gRT}{\bar{V}_1} \ln x_1$$

whereas, ideally $\pi = -\dfrac{RT}{\bar{V}_1} \ln x_1$, so that

$$g = \frac{\pi_{\text{real}}}{\pi_{\text{ideal}}} \tag{6.52}$$

It may be noted that, using relationships (6.51), the Gibbs–Duhem equation can be put

$$x_1 d\{(g - 1) \ln x_1\} + x_2 d(\ln \gamma_2) = 0 \tag{6.53}$$

A 'practical osmotic coefficient', ϕ, devised for dilute solutions, is based on the approximations: as $x_2 \to 0$, $x_1 \to 1$; $\ln x_1 = \ln(1 - x_2) \to -x_2 \to n_2/n_1$, leading to the development

$$\mu_1 = \mu_1^\circ + RT \ln a_1 \;;\; \mu_1 = \mu_1^\circ + gRT \ln x_1 \;;$$

$$\mu_1 = \mu_1^\circ - \phi RT \frac{n_2}{n_1} \tag{6.54}$$

so that
$$\phi = -\frac{n_1}{n_2} \ln a_1 \tag{6.55}$$

and the analogue of equation (6.53) is

$$d\{(1 - \phi)m\} + m d(\ln \gamma_2) = 0 \tag{6.56}$$

This scheme is suited to aqueous electrolytes because of the low molecular weight of water (for unimolal solutions, x_2 in water is $0{\cdot}0177$; in ethanol, x_2 is $0{\cdot}0441$), and the quite large deviations from ideal behaviour at considerable dilutions. For $m = 0{\cdot}5$ mole kg^{-1}, the above approximations are valid within $0{\cdot}5\%$.

6.4.2.7   *Briefly on some methods*

Of other than electrochemical methods, three are important, all based on establishment of equilibrium between solution and a separate pure phase, so that an identity of chemical potential of the component concerned is ensured. It does not matter which component is so used because if the activity coefficients of one are determined, those of the other are obtainable via the Gibbs–Duhem relation; thus, from equation (6.49)

$$\ln \gamma_2 = - \int (x_1/x_2)d(\ln \gamma_1) \tag{6.57}$$

A plot of log γ_1 against mole ratio can be integrated graphically, from a lower limit of log $\gamma_1 = 0$, to find log γ_2 values.

The basis of the freezing-point method is already to hand. Whereas in section 6.3.4 $d\mu_1 = RTd(\ln x_1)$ was equated with $\Delta S_f\, dT$ for the limiting, ideal case, for real solutions the exact equation is between $d\mu_1 = RTd(\ln a_1)$ and $\Delta S_f\, dT$. A comparison will quantify non-ideality, but it will be necessary to cater for the finite, if small, freezing-point depressions for solutions of appreciable concentration. A start can be made (reference to Fig. 6.3 will be helpful) with the differential relationship

$$-d(\ln a_1) = (\lambda/RT^2)dT \tag{6.58}$$

where λ is molar latent heat of fusion of the solvent and dT is an infinitesimal *depression* of temperature, as previously. By Gibbs–Duhem,

$$d(\ln a_2) = -(n_1/n_2)d(\ln a_1)$$

But $(n_1/n_2) = (1000/mM_1)$, where M_1 is molecular weight of solvent and m is molality of solute. Remembering the definition of the cryoscopic constant, $K_f = (RT^2/\lambda)(M_1/1000)$ (equation (6.29)), it follows that

$$d(\ln a_2) = (n_1/n_2)(\lambda/RT^2)dT = (1000\lambda/mM_1RT^2)dT = (1/mK_f)dT$$

The usual symbol for freezing-point depression is θ; since the temperature from which it is measured, T_f°, is fixed, the last equation can be written

$$d(\ln a_2) = (1/mK_f)d\theta \tag{6.59}$$

This equation is integrated[16] by use of the practical osmotic coefficient in the form

$$\phi = \theta/mK_f \tag{6.60}*$$

* To be helpful, $\phi = -\dfrac{n_1}{n_2}\ln a_1$ (equation (6.55)) $= -\dfrac{1000}{mM_1}\ln a_1$;

$$d\phi = -\frac{1000}{mM_1}d(\ln a_1) - \phi\frac{dm}{m} = d(\ln a_2) - \phi\frac{dm}{m} ;$$

$$\frac{d(\phi m)}{m} = d\phi + \phi\frac{dm}{m} = d(\ln a_2) = \frac{d\theta}{mK_f} ; \quad d(\phi m) = \frac{d\theta}{K_f}; \quad \phi = \frac{\theta}{mK_f}$$

It is convenient to define a quantity $j = 1 - \phi$ so that

$$\frac{\theta}{mK_f} = 1 - j \quad \text{and} \quad j = 1 - \frac{\theta}{mK_f}$$

Differentiation gives

$$\frac{d\theta}{mK_f} - \frac{\theta dm}{m^2 K_f} = - dj$$

or

$$\frac{d\theta}{mK_f} = (1 - j)\frac{dm}{m} - dj$$

therefore

$$d(\ln a_2) = (1 - j)d(\ln m) - dj$$

therefore

$$d\left(\ln \frac{a_2}{m}\right) = -jd(\ln m) - dj = d \ln \gamma_2$$

therefore

$$\ln \gamma_2 - \int_0^m - jd(\ln m) - j \tag{6.61}$$

which can be integrated graphically (j/m plotted against m), or by use of an empirically fitted equation. The method, capable of high accuracy, has been extensively used for electrolytes, when, evidently, m is replaced by vm. Experimentally, it requires side-by-side ice $\rightleftharpoons$ water and ice $\rightleftharpoons$ solution equilibrium systems, thermoelectric measurement of θ (at best, to 2×10^{-5} °C), and instrumental analysis of subsequently withdrawn solution. Work of this calibre calls for the dependence of latent heat on temperature to be taken into account; for this the Kirchhoff theorem is used (λ_f at 0 °C = 1436 cal mole^{-1}; C_P for liquid water and ice = 18·16 and 9·11 cal °K^{-1} mole^{-1} respectively), but to follow these matters would exceed our scope and steal thunder.[17, 18] The initial results for aqueous solutions relate to temperatures falling below 0 °C; their adaptation to other temperatures is a matter for the following subsection.

The second method prolific of accurate activity coefficient data is not, as might be supposed, ebullioscopic, because the intrinsic difficulties of this technique are too great for competitive accuracy. It is, instead, the *isopiestic*, or *isothermal distillation* method, dependent on the identity of μ_1 for solutions of the same solvent vapour pressure. Experimentally, it is simple and restful. If two solutions, not of identical vapour pressure, are contained in silver dishes nesting in hollows in a copper block, all within an evacuated, isothermal enclosure, quiet distillation occurs from one to the other and can be sensitively detected by changes in weight. It is easy to see how this can be exploited[17] by suitable interpolation of results obtained with 'unknown' solutions and standard solutions already characterised.

The third method is more specialised. It depends on the uniformity of μ_2 throughout a solid solute $\rightleftharpoons$ saturated solution system—in practice, for sparingly soluble salts. Its field of interest is the dependence of mean ion activity coefficients on *ionic strength*.

Ionic strength, I, is a misleadingly named property of electrolytic solutions, specifically defined on m or C scales by

$$I = \tfrac{1}{2} \sum m_i z_i^2 \quad \text{or} \quad I = \tfrac{1}{2} \sum C_i z_i^2 \tag{6.62}$$

where m_i or C_i is the molality or molar concentration of ion i in solution, and z_i is its charge or valency—the summation to include all kinds of ions present. Developed intuitively by Lewis and Randall in 1921, the concept that 'in dilute solutions, the activity coefficient of a given strong electrolyte is the same in all solutions of the same ionic strength'[16] received theoretical backing when Debye and Hückel published the first quantitatively successful theory of interionic attraction in 1923. Recognition of the dependence of ionic activity coefficients on ionic strength resolved some long-standing difficulties, as follows.

The Nernst solubility product and the common ion effect are encountered early in chemical studies—failures of the classical principle somewhat later. The principle is overlaid by a 'non-common ion effect', inexplicable on chemical grounds. It is that the solubility of a sparingly soluble salt is increased by addition of any more soluble salt having no ion in common.

The general explanation of this effect lies to hand in equations (6.45) to (6.47), namely,

$$\mu_{\text{salt}} = v_+\mu_+ + v_-\mu_- = \mu_{\text{salt}}^\circ + vRT \ln C + vRT \ln \gamma_\pm$$

In a saturated solution of a sparingly soluble salt, μ_{salt} is fixed by equilibrium with the solid solute; μ_{salt}° is a defined constant, so that $C\gamma_\pm = a_\pm$ is fixed by saturation at a given temperature. Accordingly, the constant expressing solubility equilibrium is a function of activities, not of concentrations. The *activity solubility product*, K_S, on the molar concentration scale, is therefore

$$K_S = (v_+ C)^{v_+}(v_- C)^{v_-} \gamma_\pm^v = (C_+ \gamma_+)^{v_+}(C_- \gamma_-)^{v_-} = a_+^{v_+} a_-^{v_-} \tag{6.63}$$

where C is the total molar concentration of the sparingly soluble salt in its saturated solution.

Suppose that to such a solid solute $\rightleftharpoons$ saturated solution equilibrium system a second, soluble, salt is added. Whatever it may be, the ionic strength is increased. All ions acquire more neighbours; interionic attraction is enhanced, and so is the non-ideality caused by such attraction. The mean activity coefficient, $\gamma_\pm$, of the ions of the sparingly soluble salt is depressed further below the infinite dilution, ideal, value of unity. If the added salt has a common ion, then, as well, C_+ or C_- is increased. The normal result is a reduction of C_- or C_+, as required to keep K_S constant. In either case, since the sparingly soluble salt must dissolve stoichiometrically, C is reduced—solubility falls. This is the common ion effect, now seen never to be considered in isolation from the influence of ionic strength. If the added salt has no common ion, it is only this influence that enters; $\gamma_\pm$ is reduced, and C must increase if equilibrium is to be maintained and K_S is to be kept constant. There is hardly any need to recall to chemically minded readers that what has been clarified is the simplest thermodynamic theory appropriate to dilute solutions.

In the hands of J. N. Brønsted and G. N. Lewis, this theory led directly to a simple and effective method of determining activity coefficients. For a sparingly soluble salt with univalent ions (e.g. TlCl), equation (6.63) simplifies to

$$K_S = C_\pm^2 \gamma_\pm^2 \tag{6.64}$$

so that

$$\gamma_\pm = (K_S)^{\frac{1}{2}}/C_\pm \tag{6.65}$$

where $C_\pm = (C_+ C_-)^{\frac{1}{2}}$, a mean ion concentration, analogous to the mean ionic functions already met. Solubilities are determined, in water, and in solutions of soluble salts without, or with a common ion. Log $C_\pm$ is linear, or tends to linearity, with $I^{\frac{1}{2}}$; extrapolation to $I = 0$ ($\gamma_\pm \to 1$) yields log $(k_s)^{\frac{1}{2}}$ With K_S so determined, $\gamma_\pm$ at any ionic strength is calculable from equation (6.65).

In a celebrated example of such work, Brønsted and la Mer[19] chose sparingly soluble salts ('solute salts') derived from ions such as

$$[Co(NH_3)_4.C_2O_4]^+, \ [Co(NH_3)_6]^{3+}, \ [Co(NH_3)_2(NO_2)_2.C_2O_4]^-, \ S_2O_6{}^{2-}.$$

The reasons for such a choice were that these salts, of alternative valency types (1:1, 1:2, 3:1), are well-defined and crystalline, with suitable solubilities ($\sim 10^{-4}$ mole l^{-1} at 15 °C), lending themselves to a common analytical estimation method (ammonia distillation). More important, they have large ions, a factor foreseen to minimise *specific* interionic actions. A variety of added salts ('solvent salts'; NaCl, KNO₃, BaCl₂) provided a range of ionic strengths. The results gave linear plots of the kind already indicated, with slopes, in relation to valency type, as required by the (then new) Debye–Hückel theory.

More than this *tour de force* came out of Brønsted's work. It had already led him to his *principle of the specific interaction of ions*, which is, in a sense, the chemical complement of the purely physical Debye–Hückel theory. The thinking behind it is that a 'Debye–Hückel model', of charged bodies exerting long-range electrostatic forces on each other, is too simple to deal with ionic non-ideality, except for solutions of extreme dilution. Ions are chemical entities, prone to short-range, specific interactions. Yet such interactions must fall under the control of the longer-range coulombic forces. Forces of attraction and repulsion are respectively strengthened and weakened by the responses they evoke. Hence, in a solution, ions of opposite charge will often come close to each other, whereas ions of the same charge will rarely do so. It follows that only specific interactions between *oppositely* charged ions are likely to be significant, and it is only here that effects due to ionic size, shape, polarisability and the like need be anticipated. Using the analogy of a mixture of imperfect gases (section 6.4.2.2) only the virial coefficient B_{12} is significant; B_{11} and B_{22} are not. This masterly example of intuitive chemical reasoning has stood the test of time.

6.4.2.8   *Some thermal properties of solutions*

It is recalled that studies of systems conducted over a range of temperatures open up a wider field of information than those restricted to a single temperature. This may be especially so for liquid mixture and solution systems, where sensitivity of the energy–entropy balance is to be expected, and, perhaps, a strong dependence of non-ideality on temperature. How this is best examined depends on the nature of the system concerned.

It may be helpful to maintain the view that μ_i remains the centre of interest, and to write out again the familiar equation

$$\left(\frac{\partial G}{\partial n_i}\right)_{T,P,n_j} = \bar{G}_i = \mu_i = \mu_i^\circ + RT \ln a_i = \mu_i^\circ + RT \ln x_i + RT \ln \gamma_i$$

To consider the dependence of μ_i on temperature, we look first at the right-hand side of the final equality. The first term, μ_i°, is a constant, independent of the mole fraction, x_i, of component i, but dependent on temperature; $(\partial \mu_i^\circ / \partial T)_P = -\bar{S}_i^\circ$. In so far as the standard state is an ideal state, this has nothing to do with non-ideality. The second term, $RT \ln x_i$, is a function only of the temperature and the ideal entropy of mixing (cf. equations (6.24) to (6.26)), which is not temperature dependent. This term, linear with temperature, is then a purely idealised, entropic term, and also has nothing to do with non-ideality. If the system were ideal, these first two terms would be enough. The third term, $RT \ln \gamma_i$, includes all the entropic and energetic contributions to μ_i arising from non-ideality.

To complete the revisionary inspection of terms, it is noted that the activity of component i, $a_i = \gamma_i x_i$, incorporates both ideal entropic and non-ideality terms. The activity in the standard state, a_i°, is clearly always unity, and, by definition, has no dependence on temperature. Only activities in non-standard states, a_i, have such dependence.

The last two paragraphs could be rewritten in terms of m_i instead of x_i; for dilute solutions, the one is proportional to the other, but the essential likeness is that both are independent of temperature. This is not true for the molar concentration of a solution.

It is advantageous to remove the linear contribution to temperature dependence, and write—in the context of solutions

$$\mu_i/T = (\mu_i^\circ/T) + R \ln m_i + R \ln \gamma_i \tag{6.66}$$

Then, observing that the middle term on the right-hand side of this equation is not temperature dependent, and recollecting that partial molar quantities are related together in the same way as molar quantities (section 6.2.2), it is seen that

$$\left\{ \frac{\partial}{\partial T} \left(\frac{\mu_i}{T} \right) \right\}_P = - \frac{\bar{H}_i}{T^2} \tag{6.67}$$

where $\bar{H}_i$ is the partial molar enthalpy of component i, we can write

$$\left\{ \frac{\partial}{\partial T} \left(\frac{\mu_i}{T} \right) \right\}_P \Big/ \left\{ \frac{\partial}{\partial T} \left(\frac{\mu_i^\circ}{T} \right) \right\}_P + R \left(\frac{\partial \ln \gamma_i}{\partial T} \right)_P$$

therefore

$$- \frac{\bar{H}_i}{T^2} = - \frac{\bar{H}_i^\circ}{T^2} + R \left(\frac{\partial \ln \gamma_i}{\partial T} \right)_P$$

where $\bar{H}_i^\circ$ is the partial molar enthalpy of component i in its standard state. Then

$$\left(\frac{\partial \ln \gamma_i}{\partial T} \right)_P = \frac{\bar{H}_i^\circ - \bar{H}_i}{RT^2} \tag{6.68}$$

Clearly, also

$$\left(\frac{\partial \ln a_i}{\partial T} \right)_P = \frac{\bar{H}_i^\circ - \bar{H}_i}{RT^2} \tag{6.69}$$

Care is necessary in relation to standard states.

For a problem concerned with a liquid mixture of two equivalent components, symmetrical activity coefficients are used, and the standard states are those of the pure components. Then, $\bar{H}_1^\circ = H_1^\circ$ and $\bar{H}_2^\circ = H_2^\circ$, but it is better to use the second of each pair of symbols (i.e. without the 'partial' bar) as a reminder.

For a solution problem, the standard state for the solvent is that of the pure solvent; without exception, $\bar{H}_1^\circ = H_1^\circ$, and either symbol can be used without ambiguity. But unsymmetrical activity coefficients would be in use for the purpose of studying the Henry's law non-ideality of the solute *in solution*. The standard state for the solute must be one in which infinite dilution conditions prevail; it can be, for example, the 'hypothetically ideal unimolal solution', but since H (unlike G) has nothing to do with S, neither m_2 nor x_2 are relevant, and the standard state can be equally quoted as that of infinite dilution. In either case, $\bar{H}_2^\circ$, the partial molal enthalpy of the solute at infinite dilution must be used, and is not to be identified, or confused, with H_2°.

Since no absolute values can be assigned to the partial molar enthalpies of substances in solution, it is convenient to define a *relative partial molar enthalpy*, symbolised $\bar{L}_i$:

$$\bar{L}_i = \bar{H}_i - \bar{H}_i^\circ \qquad (6.70)$$

where, of course, the 'bar' must be retained for generality. Then equations (6.68) and (6.69) become, for the solvent

$$\left(\frac{\partial \ln \gamma_1}{\partial T}\right)_P = \left(\frac{\partial \ln a_1}{\partial T}\right)_P = -\frac{\bar{L}_1}{RT^2} \qquad (6.71)$$

and for solute,

$$\left(\frac{\partial \ln \gamma_2}{\partial T}\right)_P = \left(\frac{\partial \ln a_2}{\partial T}\right)_P = -\frac{\bar{L}_2}{RT^2} \qquad (6.72)$$

Similarly, *relative partial molar heat capacities* at constant pressure can be defined for either component; they are symbolised $\bar{J}_i$, so that

$$\bar{J}_i = \bar{C}_{P_i} - \bar{C}_{P_i}^\circ \qquad (6.73)$$

and it is clear that

$$\bar{J}_i = \left(\frac{\partial \bar{L}_i}{\partial T}\right)_P \qquad (6.74)$$

For an ideal solution, $\bar{L}_i = \bar{J}_i = 0$. It is therefore evident that, in principle, the study of the temperature dependence of activity coefficients opens up a rich field of information—$\bar{L}_i, \bar{J}_i$ and (as can be seen without detailed comment) $\bar{S}_i$—which might be expected to tell us a great deal about the non-ideal interactions in solutions. This is what we need to understand them better. Unfortunately, there is a snag, encountered before. Consider that the route to the information starts with a *deviation*—usually a difference between larger quantities. This has to be differentiated with respect to temperature, incurring a heavy loss of accuracy. A second operation of this kind may have small chance of retaining significance unless the primary measurements are very accurate. How far this will be disabling will vary from one case to another, but, in principle, it would seem better to get at thermal quantities directly by using a calorimeter.

Unfortunately again, calorimetry has its own difficulties, especially in relation to dilute solutions. Yet it is for such solutions that accurate thermal data are very desirable because they present the best chance of theoretical interpretation. For electrolytic solutions, for example, developments of the Debye–Hückel theory (well tested in some connections, but not invulnerable) have been of great assistance. By and large, all the available methods are applied in this field, with some overlap and interagreement. It may be noted, avoiding details of method,[17] that equations (6.72) and (6.74) are applied, using calorimetric data, to 'converting' freezing-point activity coefficients to higher temperatures.

In this field of calorimetry there is greater difficulty in determination of differential than of integral thermal quantities. The former can be derived from the latter by methods similar to those normally used to evaluate partial molar properties, but these need not be discussed. For this reason, most, but not all, accurate data come from experimental *total* or *integral heats of solution* and *dilution*. The first of these is generally understood to mean ΔH for the solution of one mole of solute, initially in its normal, pure, standard state, either in a specified number of moles of solvent, or in solvent to give a solution of stated molality. It is a composite quantity, important in thermochemistry in a manner not yet discussed, incorporating (for solid solutes) contributions from lattice energy and (in general) solvation. Heat of solution is a function of the composition of the solution formed because of a finite heat of dilution, which, ideally, would be zero. For electrolyte solutions, its main origin can be seen in the energy required to separate ions against a net force of interionic attraction. In all cases ΔH_{soln} tends asymptotically, with increasing dilution of the solution formed, to a 'heat of solution to infinite dilution'. This forms a link between the alternative standard states and provides the quantitative difference between

H_2° and $\bar{H}_2^\circ$—a distinction discussed above. The integral heat of dilution is the difference between any two integral heats of solution for a given solute in a given solvent, i.e. it is ΔH per mole of solute for a dilution from m_1 to m_2. For given m_1, it tends asymptotically to a constant value as m_2 tends to zero,

The general distinction between integral and differential heats is clear from the following self-explanatory equations:

$$\Delta H = n_1\bar{H}_1 + n_2\bar{H}_2 - n_1 H_1^\circ - n_2 H_2^\circ \tag{6.75}$$

$$= n_1(\bar{H}_1 - H_1^\circ) + n_2(\bar{H}_2 - H_2^\circ) \tag{6.76}$$

$(\bar{H}_2 - H_2^\circ)$ is the *partial or differential molar heat of solution*; $(\bar{H}_1 - H_1^\circ)$ is called the *partial or differential molar heat of dilution*. There is some danger of confusion in relation to dilution (per mole of solute, or per mole of solvent?); the name of 'partial or differential molar heat of solution of the solvent' has accordingly been suggested for $(\bar{H}_1 - H_1^\circ)$.

There are two reasons for inviting attention to the thermal data for aqueous sodium chloride solutions entered in Table 6.2.

Table 6.2   Some thermal properties of aqueous NaCl solutions at 25 °C *

m_2	$\bar{L}_1$	$\bar{L}_2$	$\bar{C}_{P_1}$	$\bar{C}_{P_2}$
	cal mole^{-1}		cal °K^{-1} mole^{-1}	
0	0	0	17·9956	−22·1
0·2	0·1	+80	17·984	−14·20
0·5	0·8	0	17·951	−8·65
1·0	3·3	−220	17·866	−2·30
2·0	10·0	−445	17·621	+6·90
4·0	21·5	−688		
6·0	12·3	−595		

* Interpolated from data cited in ref. 18.

The first reason is to offer a reminder of the way in which partial molar extensive properties must be thought about. If this is forgotten, the large negative $\bar{C}_{P_2}$ for NaCl at infinite dilution in water comes as a shock; what possible significance can be attached to a negative heat capacity—does the temperature fall when heat is supplied? This is, of course, a nonsense. The negative $\bar{C}_{P_2}$ means that for an infinitesimal addition, dn_2 mole, of solute, the heat capacity of *the whole solution system* is reduced in its positive value by dC_P, and, proportionally, per mole of solute, the reduction tends at infinite dilution to 22·1 cal °K^{-1}. The second reason is merely to indicate the kind of thinking to which this leads. How is this effect to be explained?

The first essential to note is that, although the name and symbol are attached to the solute, the origin of the effect is to be looked for in the solvent—in terms of an action of the solute on it. It may be remembered that the heat capacity of ice is about half that of water. This leads to the conjecture that the solute particles have a local 'freezing-up' effect on the water round them. Why should they have such an effect? In this case, because they are ions, providing an intense, inhomogeneous electric field to which the polar and polarisable solvent molecules are unlikely to be indifferent. In short, it is tolerably certain that we are looking at a thermal consequence of ionic hydration. If this is so, the figures suggest that these effects are rather strongly dependent on concentration, so perhaps a simple freezing-up effect is not enough, and we shall have to look for something co-operative. Of course, a freezing-up can act both ways—if it is not, all of it, very strong, it might be vulnerable to 'melting', and this might increase $\bar{C}_{P_2}$ instead of reducing it. The complex nature of C_P has already been indicated in earlier discussion. Perhaps we had better look at the partial molar entropies as well. But, no, this is enough. Two general points have been made. Two other sober thoughts may be added, as follows.

Whatever the complexities, the columns relating to the two components in Table 6.2 must accurately conform to the Gibbs–Duhem equation. The second thought is that here is an example, as mentioned in chapter 1, of thermodynamic data shouting that something needs to be investigated. It would be better than shouting if there were more accurate data available over both a range of temperatures and a range of pressures (since solute–solvent interactions involve significant volume changes). As it stands, an enormous field of work essential to understanding of solution systems awaits attention.

6.5   Thermodynamic excess functions

Liquid mixtures of two components show, over the whole composition range, a wider sweep of non-ideality, best surveyed in terms of 'excess functions'. It is recalled that, for the formation of a mole of mixture,

$$\Delta G^M = \Delta H^M - T\Delta S^M \tag{6.23}$$

The differences between the extensive terms in this equation and those relating to ideal mixing are the excess functions. Since

$$\Delta S^M_{ideal} = -R(x_1 \ln x_1 + x_2 \ln x_2) \tag{6.1}$$
$$\Delta G^M_{ideal} = RT(x_1 \ln x_1 + x_2 \ln x_2) \tag{6.24}$$
$$\Delta H^M_{ideal} = 0$$

then, for example

$$\Delta G^E = \Delta G^M - \Delta G^M_{ideal} \qquad (6.77)$$

$\Delta H^E = \Delta H^M$ can, with ΔS^E, be obtained from the temperature dependence of ΔG^E, but usually better from calorimetric heats of mixing.

Two systems previously considered (cf. Fig. 6.4) are represented in this alternative way in Fig. 6.7.

The first of them, acetone–chloroform, shows the expected negative ΔG^E, indeed, calculated from the negative deviations from Raoult's law. Exothermic mixing is consistent with the earlier suggestion of hydrogen-bonding between the components—although maximum heat liberation does not occur at the 1:1 molar proportion. It is also intelligible that the supposed bonding calls for a degree of intercomponent organisation that exacts a penalty; $T\Delta S^E$ is negative and large enough to 'use up' most of the negative ΔH^M, leaving ΔG^E much less negative than it might otherwise be.

The second system, water-t-butyl alcohol, is less easy. Of course ΔG^E is large and positive, but the nearly symmetrical $\Delta G^E(x_2)$ plot—derived from isothermal measurements—itself gives no clue to the astonishingly unsymmetrical behaviour of ΔH^M and $T\Delta S^E$ as functions of x_2. It did not escape notice when measurements were extended to a range of temperatures. However determined, these thermal quantities bring in more information than available from the free energy term alone.

A first look at this purely thermodynamic evidence shows that this system raises not one but several problems. The positive deviations from Raoult's law at low x_2 are evidently due to negative $T\Delta S^E$—in itself suggesting a considerable degree of 'ordering'. On the other hand, at high x_2, the deviations have an energetic origin. Why is it that mixing is exothermic over one half of the composition range, and endothermic over the other half? How is it that, with water in excess, large, *positive* deviations from Raoult's law are associated with *heat liberation* on mixing—and also with loss in volume of the alcohol (cf. Fig. 6.2; the effect is greater for t-BuOH than for EtOH)?

Comparisons are helpful. Water and dioxan mixtures behave almost identically. Within a series of monohydric alcohols mixed with water, the asymmetry in ΔH^M, almost absent for methanol, increases with molecular weight, but, for a given alcohol, decreases with rising temperature. This effect, together with a very steep dive of ΔS^E to negative values at low x_2, seems to be a function of the hydrocarbon part of an otherwise hydrophilic molecule. Hydrocarbons themselves dissolve in water with *liberation of heat*, but with so large a *negative* ΔS^E as to keep solubility very low. They also form crystalline hydrates, which have been known to block natural

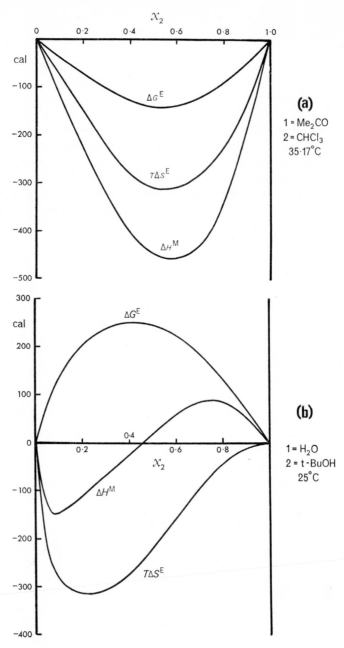

Fig. 6.7   Thermodynamic excess functions of mixing for
(a) acetone and chloroform at 35·17 °C.
(b) water and t-butyl alcohol at 25 °C.

gas pipe–lines. All this, and much else, has led to acceptance of 'hydrophobic hydration'. It is ascribed to a structure-stabilising effect of a 'hydrophobic' molecule, or part of a molecule, on the water surrounding it. There is a local 'freezing effect', quite different from that induced by ionic solutes, reflected in negative ΔH^M and $T\Delta S^E$.

This goes some way to explain Fig. 6.7b. At low x_2, there is exothermic structure-making; at high x_2, endothermic structure-breaking—crudely, depolymerisation of the water. There is a good deal more to this complex system, attacked, or under attack, by dielectric relaxation, ultrasound absorption, proton magnetic resonance and cold neutron inelastic scattering methods. This brief discussion, with items in sections 6.2.3 and 6.4.1 completes the little scheme of correlated 'interludes', designed to show how clearly thermodynamics poses problems of major interest—this is the main objective, with encouragement of more specific reading elsewhere.[4,20]

6.6   Miscibility gaps

Partial miscibility is familiar in binary liquid systems. For example, phenol and water form a complete range of homogeneous mixtures above an *upper critical solution temperature* (UCST) of 65·85 °C, but not below, so that, on cooling, certain homogeneous mixtures of these components separate into two liquid layers—mutually saturated *conjugate solutions*. It is instructive to view such behaviour in terms of activity coefficients, and of the thermodynamic functions of mixing.

Raoult's law activity coefficients for the components of a binary mixture can be expressed to any order of accuracy by empirical polynomial equations such as

$$\ln \gamma_1 = A_1 x_2 + B_1 x_2^2 + C_1 x_2^3 \dots$$

$$\ln \gamma_2 = A_2 x_1 + B_2 x_1^2 + C_2 x_1^3 \dots$$

$$(6.78)$$

where A_1, B_1 ..., A_2, B_2 ... are numerical coefficients. No matter how extreme deviations from ideal behaviour may be, the Gibbs–Duhem equation must be obeyed. This means that the coefficients A_1, A_2; B_1, B_2 ... in the two equations are not independent. To find how they must be related, $x_2(d(\ln \gamma_2)/dx_2)$ and $(1-x_2)(d(\ln \gamma_1)/dx_2)$ are evaluated. They must add to zero (cf. equation (6.49)). Coefficients of like powers of x_2 must also add to zero; this provides equations soluble for the desired relations. The most important is

$$A_1 = A_2 = 0 \tag{6.79}$$

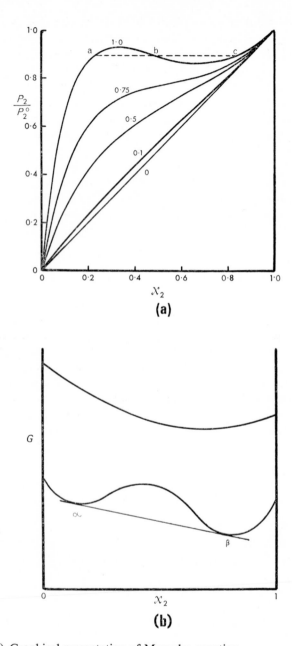

Fig. 6.8   (a) Graphical presentation of Margules equation
(b) The types of $G(x_2)$ relation for complete and partial miscibility.

and this leads, with neglect of higher powers of x_2, to the *Margules equations*:

$$\ln \gamma_1 = \alpha x_2^2 \; ; \; \ln \gamma_2 = \alpha x_1^2 \tag{6.80}$$

or $\quad P_1 = x_1 P_1^\circ \exp(\alpha x_2^2) \; ; \; P_2 = x_2 P_2^\circ \exp(\alpha x_1^2)$ $\qquad$ (6.81)

where α is a single parameter, zero for ideality. This turns out to be a remarkably satisfactory approximation. Figure 6.8a shows plots of $P_2/P_2^\circ(x_2)$ for $\alpha' = \alpha/\ln 10$ increasing from 0 to 1, conforming to Henry's and Raoult's laws at low and high x_2, respectively. The uppermost curve, however, has developed a maximum and a minimum, implying (as indicated by the broken, horizontal line) that there might be three liquid mixtures, a, b and c, of identical P_2 at the single temperature concerned—and, of course, similarly for the other component. But for a two-component system the phase rule, *at fixed temperature* $(F = C - P + 1)$, limits the number of phases that can coexist to three—in this case, two liquid, one vapour. The curve between a and c is imaginary (cf. van der Waals and Andrews isothermals) and the real P_2/P_2° plot is indeed horizontal between a and c. This is the miscibility gap; a and c are the compositions of the two liquid layers in mutual equilibrium. Over this range, P_1 and P_2 are fixed, and so are the total pressure and composition of the vapour.

Let us next consider the total Gibbs free energy, G, of $n_1 + n_2 = 1$ mole of two components, expressing over-all composition in terms of x_2. For the existence of a complete range of homogeneous mixtures, the $G(x_2)$ plot must everywhere be concave upwards, as in the upper curve of Fig. 6.8b. A contrary case is imaginable. If ΔH^M were sufficiently positive, then, remembering the dominant effect of ΔS^M at low and high x_2, $G(x_2)$ might be calculated as in the lower curve of Fig. 6.8b, but it would be in part imaginary, not representing real behaviour. To find why, we ask the question—always apposite—'is there any way in which this system could decrease its total G?' The answer is 'yes'. All hypothetical systems represented by points along the continuous curve lying above a tangent touching the curve at α and β, near the two minima, can decrease in G by separating into distinct liquid phases of x_2 respectively equal to the abscissae of α and β. These are the conjugate solutions. They are, in effect, two subjoined liquid systems making their proportional, additive contributions to the sum-total G. In other words, the sum-total G for all the two-phase systems within the miscibility gap lies along the tangent. It is to be seen that these liquid phases (call them α and β) have identical $(\partial G/\partial x_2) = -(\partial G/\partial x_1)$.

This is consistent with the requirement for equilibrium that

$$\mu_1^\alpha = \mu_1^\beta; \; \mu_2^\alpha = \mu_2^\beta \qquad (6.82)*$$

Of course such equalities cannot extend to $\bar{H}_i$ and $\bar{S}_i$, otherwise the phases could not differ. It is to be expected, for example, that $\bar{H}_1^\beta > \bar{H}_1^\alpha$ and $\bar{S}_1^\beta > \bar{S}_1^\alpha$, i.e. the enthalpy and entropy of a component should be greater in the phase in which it is the more dilute. Then transfer of component 1 from phase α to phase β (or of component 2 in the reverse direction) would occur with absorption of a latent heat, $\Delta H = T\Delta S$, very like the evaporation of a pure liquid into its saturated vapour. This analogy can be pressed further—aided by recollection of the Andrews isothermals embracing liquefaction. Compositions and properties of the phases in equilibrium come closer with rising temperature until, at a critical temperature, they coincide. There is indeed much in common between the two kinds of critical phenomena, even to the extent of evidence for 'clustering' in one-phase systems in the *critical region*—for the solution systems, 'critical opalescence', anomalous viscosity and sound absorption.

---

* Not obvious because $dG/dx_2 \neq (\partial G/\partial n_2)_{n_1}$. To prove:

$$G = x_1\mu_1 + x_2\mu_2 \; ; \quad \frac{dG}{dx_2} = x_1\frac{d\mu_1}{dx_2} - \mu_1 + x_2\frac{d\mu_2}{dx_2} + \mu_2$$

(N.B., $\dfrac{dx_1}{dx_2} = -1$)

By Gibbs–Duhem,

$$x_1\frac{d\mu_1}{dx_2} + x_2\frac{d\mu_2}{dx_2} = 0 \; ; \quad \frac{dG}{dx_2} = \mu_2 - \mu_1$$

$$G = x_1\mu_1 + x_2\mu_2 + x_2\mu_1 - x_2\mu_1$$

$$= (x_1 + x_2)\mu_1 + x_2(\mu_2 - \mu_1) = \mu_1 + x_2\frac{dG}{dx_2}$$

For the two phases, α and β, with common dG/dx_2 ,

$$\mu_1^\beta = G^\beta - x_2^\beta\frac{dG}{dx_2} \; ; \quad \mu_1^\alpha = G^\alpha - x_2^\alpha\frac{dG}{dx_2}$$

therefore

$$\mu_1^\beta - \mu_1^\alpha = G^\beta - G^\alpha - (x_2^\beta - x_2^\alpha)\frac{dG}{dx_2} .$$

But for the tangent,

$$\frac{dG}{dx_2} = (G^\beta - G^\alpha)/(x_2^\beta - x_2^\alpha)$$

therefore

$$\mu_1^\beta - \mu_1^\alpha = 0 \quad \text{Similarly, } \mu_2^\beta - \mu_2^\alpha = 0$$

There is no great difficulty in understanding UCST. It is the temperature at which thermal motion enforces mixing, and is reached when $\frac{1}{2}RT$ becomes commensurate with the unfavourable, positive ΔG^E. Entropy must increase with rising temperature, and eventually prevails over the energy term associated with the intracomponent forces tending to maintain segregation. It is reasonable to class the UCST as an 'entropy effect'.

Binary liquid systems with a *lower critical solution temperature* (LCST) present more difficulty—and interest. Some of these have an observable UCST as well, so that there is a closed solubility loop. Within the loop, plotted between T and x_2 axes, there are two liquid phases in equilibrium; outside it, there is one homogeneous phase. Some preliminary implications can be drawn, as follows.

For such a system, cooled progressively from above the UCST to below the LCST, the $G(x_2)$ plot must change oddly—in terms of the types of plot in Fig. 6.8b, from upper to lower, then back to upper again. The contributions to G must change radically in relative significance. Above the 'waist' of the loop, mutual solubility increases with rising temperature, so the solution process must be endothermic.* Below the waist, it is the reverse, so that the solution of one phase in the other is exothermic. It is evident that the nature of the homogeneous phases separated by the miscibility gap must be changing remarkably. Perhaps the most striking conclusion comes by following up the basic fact that the entropy of a system must decrease with falling temperature. The observation of two liquid phases spontaneously mixing *on cooling* is extraordinary, because entropy of mixing is always positive. There must be some larger, but less apparent, compensating entropy loss—some higher degree of order must be established in the homogeneous phase produced.

Some sort of explanation comes from considering the nature of the earlier and better-known systems showing this effect. One of the components is water (light or heavy), except it is glycol or glycerol in a few instances. The other component is an organic substance, 'in the middle of a homologous series', having 'bifunctional' molecules—each with a hydrophilic, water-solubilising group attached to a hydrocarbon residue (di- and triethylamines, glycol ethers, nicotine, etc.). Then, classically, the explanation was 'compound formation'—later translated as intercomponent hydrogen-bonding. Not strong enough to override the incompatibility of the components above the LCST, it just succeeds in pulling them into complete admixture below it. On this basis, the LCST has sometimes been

* This is seen at once by application of the le Chatelier principle, assumed to be well-known to all readers. Since solubility is of the nature of an equilibrium constant, the van't Hoff equation applies.

called an 'energy effect', but this is an oversimplification neglecting proper identification of the accompanying large entropy loss.

It is true that hydrogen-bonding, requiring special mutual location and orientation of bonded molecules, would cause loss of translational and rotational freedom, but at least one, if not both of the components are extensively hydrogen-bonded before mixing. It is satisfying to picture the unlike molecules, forced into independent rotation with rising temperature, no longer being able to present attractive faces to each other, but only averaged-out 'repulsive' aspects. Doubt has been expressed whether hydrogen-bonding is enough. Below the LCST, there is always a large, negative volume of mixing. Deuterium oxide and α-picoline (LCST, 92 °C; UCST, 112·5 °C) show no miscibility gap at pressures above 90 atm. All the systems concerned are of the type discussed in the preceding section. Thought is invited.

There is another kind of LCST not infrequently encountered in solutions of high polymers, e.g. polystyrene in cyclohexane (UCST, 30 °C; LCST, 180 °C), polyisobutane in benzene (UCST, 23 °C; LCST, 160 °C). Where there is an *upper* CST as well, it is at a *lower* temperature than the lower CST. As an aid to thinking this out, a single-phase system lies between two two-phase systems on the temperature scale—instead of a two-phase system lying between two single-phase systems. This behaviour has recently been found for methane and hex-1-ene.[21] It seems to depend on large disparities between the components in molecular weight and coefficient of expansion; the LCST lies not far below the normal liquid–gas critical temperature of the more volatile component, where the free volume is very large. Mixing not far below the LCST is typified by large negative ΔH^M, ΔS^M and ΔV^M, perhaps intelligible as results of mixing components, one very 'open', the other compact. This is as far as we can proceed towards an 'explanation', but it may be useful to point again to the great spread in properties of many pure liquid phases between T_M and T_c. Sulphur-hydrocarbon systems, showing high temperature LCST, provide an almost grotesque example, but here it is polymerisation of the molten sulphur with rising temperature that is responsible for a drastic change of state of one of the components. It must be noted that the field of liquids and liquid mixtures requires its own dedicated study[1,22,23]—a fact as clearly demonstrated by CST phenomena (a critical compilation of data is available[24]) as by others.

6.7 Briefly on some other potentials

Some account must be given of systems not adequately defined in state by specification of T, P, Σn_1. This calls for an excursion into allied fields of surface chemistry and electrochemistry—important enough in the scheme of things.

6.7.1   'Surface tension'

Atoms, ions or molecules in the surface of a phase differ in environment from those in the bulk of the phase. They have taken a step towards evaporation. They have fewer nearest neighbours, unsymmetrically placed, and this is bound to cause local differences in spacing and other aspects of three-dimensional geometry. For ionic crystals, planes containing positive and negative ions, normally coincident, may become separated, giving a dipolar surface. For metals, there may be 'electron overlap'. Non-spherical or polar molecules of liquids will assume preferred orientations in the surface layer. In general, there will be a dissymmetry of unsatisfied fields of force, abhorrent to Nature. Whatever can happen to relieve this stress will do so—reduction of surface area, as permitted, by liquids, or the acquisition by surface entities of unlike (as next best to like) 'outer partners'. This is why *adsorption* is ubiquitous.

Thermodynamics deals with macroscopic systems in equilibrium. 'Surface systems' are macroscopic in only two dimensions, and are seen never to be 'in equilibrium' in the normal sense. How is thermodynamic reasoning to be applied in these circumstances?

Consider the tranquil surface of a liquid in contact with its own saturated vapour in an enclosure. The system is stable, and does not change with time. There is exchange of molecules between liquid and vapour phases, which are undoubtedly in equilibrium with each other. We have learned that 'equilibrium' is relative, so why should we not consider the whole system, liquid–vapour interface included, to be well and truly in equilibrium for the purposes of thermodynamic argument? But we cannot be entirely happy about this as it stands, for the following reason.

It is a natural process for liquids to decrease their surface area, assuming the spherical shape of least possible ratio of surface area to volume—$1/3r$, where r is radius. If r is very small, and the ratio large, the vapour pressure of the liquid is increased. Small droplets are unstable with respect to large ones; supersaturation is necessary for a vapour to condense to droplets. So 'surface liquid' is in a higher free energy state than 'bulk liquid', and the generation of surface must require a positive ΔG, i.e. it calls for expenditure of work. It is obvious that this must be so, because cohesive attractions must be broken in the tearing away of some nearest neighbours, but this is not quite the point—it is that the ΔH and $T\Delta S$ of surface formation do not 'compensate', leaving ΔG zero, as required for equilibrium, and as for complete transfer of substance from liquid to saturated vapour.

Thermodynamics therefore requires an intensive quantity to deal with the stress in a surface, complementary to the extensive quantity, surface area, and an appropriate term must be added to each of the four fundamental Gibbs equations. In place of equation (6.18), we must write

$$dG = -SdT + VdP + \sum \mu_i dn_i + \gamma dA \qquad (6.83)$$

where γ is *surface tension*, $(\partial G/\partial A)_{T,P,ni}$, and A is area of surface. This recognises that the total G of, for example, our enclosed liquid–vapour system is a function of the area of surface it contains. It is to be noted that surface tension (dyne cm^{-1}) is dimensionally the same, and is better regarded, as *surface free energy* (erg cm^{-2}). Its temperature coefficient at constant pressure, with sign reversed, is the entropy of formation of unit area of surface.*

Equation (6.83) holds generally for any interface in a polycomponent system, and is now to be used to derive the *Gibbs adsorption equation*.

Consider two homogeneous phases, α and β, containing any number of components, in equilibrium with each other across a flat interface—flat because across a spherical surface of radius r, having a surface tension γ, there is a pressure difference $2\gamma/r$. We have no information about special conditions at or near the interface—in particular to what extent the distribution of the components may be abnormal. We therefore place two imaginary planes, one on either side of the interface, far enough apart to ensure that they enclose all regions in which there is any inhomogeneity. This enclosure, containing the interface, and as much or more than necessary of the phases α and β, is called the *interphase*, δ.

Suppose that this system, having no tendency to change, and with equilibrium distribution of the components throughout all its parts, is subjected to an infinitesimal test process, conducted at constant temperature, pressure and material content. The process is to increase the area of the interphase by dA^σ—this can, of course, be effected by change of shape. Adapting equation (6.83),

$$dG = \sum(\mu_i^\sigma dn_i^\sigma + \mu_i^\alpha dn_i^\alpha + \mu_i^\beta dn_i^\beta) + \gamma dA^\sigma \qquad (6.84)$$

Since material content is fixed, $dn_i^\sigma = -(dn_i^\alpha + dn_i^\beta)$. This is consistent with unimpeded interchange, as already postulated, of all components between bulk phases and interphase, so that the distribution of components will be such as to minimise G.

---

* There is a difficulty here. For the one-component, liquid-saturated vapour system, P is the equilibrium vapour pressure, and cannot be kept constant when temperature is varied. This is why constant volume conditions are often adopted in discussion, but not at present—to avoid clouding issues otherwise uncomplicated.

Accordingly, we must write

$$\mu_i^\sigma = \mu_i^\alpha = \mu_i^\beta = \mu_i \tag{6.85}$$

discarding unnecessary superscripts. This brings the term in brackets in equation (6.84) to zero. Hence, for the test process,

$$(dG)_{T,P,n_i} = \gamma dA^\sigma \neq 0 \tag{6.86}$$

which agrees with the definition of γ and does not contradict the preceding discussion.

It is, however, desired to concentrate attention on the interphase. Rearranging equation (6.84),

$$
\begin{aligned}
dG &= (\gamma dA^\sigma + \Sigma\mu_i^\sigma dn_i^\sigma) + \mu_i^\alpha dn_i^\alpha + \mu_i^\beta dn_i^\beta \\
&= \qquad dG^\sigma \qquad\quad + (dG^\alpha + dG^\beta)
\end{aligned}
$$

This involves treating the interphase as a thermodynamic system and defining dG^σ as

$$dG^\sigma = \gamma dA^\sigma + \Sigma\mu_i^\sigma \, dn_i^\sigma \tag{6.87}$$

so that μ_i^σ itself requires definition as

$$\mu_i^\sigma = \left(\frac{\partial G^\sigma}{\partial n_i^\sigma}\right)_{T,P,n^\sigma_j,A^\sigma} \tag{6.88}$$

Integration of equation (6.87) with all intensive quantities kept constant gives the self-evident result

$$G^\sigma = \gamma A^\sigma + \Sigma\mu_i^\sigma n_i^\sigma \tag{6.89}$$

General differentiation leads to

$$dG^\sigma = \gamma dA^\sigma + A^\sigma d\gamma + \Sigma\mu_i^\sigma dn_i^\sigma + \Sigma n_i^\sigma d\mu_i^\sigma \tag{6.90}$$

and, from comparison with (6.87),

$$A^\sigma d\gamma + \Sigma n_i^\sigma d\mu_i^\sigma = 0 \tag{6.91}$$

which is the analogue of the Gibbs–Duhem equation for an interphase. If, because $\mu_i^\sigma = \mu_i$, we drop the superscript to μ_i, this becomes for a two-component system

$$A^\sigma d\gamma + n_1^\sigma d\mu_1 + n_2^\sigma d\mu_2 = 0 \tag{6.92}$$

which is suited for application to the free surface of a solution of a single solute. Confining further attention to this case, the next step is to consider that the Gibbs–Duhem equation must apply to the bulk phase of the solution, so that

$$n_1^{\circ} d\mu_1 + n_2^{\circ} d\mu_2 \tag{6.93}$$

where the superscript is used to indicate 'bulk phase'.
Multiplication by n_1^{σ}/n_1° gives

$$n_1^{\sigma} d\mu_1 + \frac{n_1^{\sigma} n_2^{\circ}}{n_1^{\circ}} d\mu_2 = 0$$

Subtraction from equation (6.92) leads to

$$A^{\sigma} d\gamma + \left(n_2^{\sigma} - \frac{n_1^{\sigma}}{n_1^{\circ}} \cdot n_2^{\circ} \right) d\mu_2 = 0 \tag{6.94}$$

The interphase is made up of n_2^{σ} mole of solute associated with n_1^{σ} mole of solvent. In the bulk solution phase, however, $(n_1^{\sigma}/n_1^{\circ}) \, n_2^{\circ}$ mole of solute is associated with n_1^{σ} mole of solvent. Then a *surface excess* of solute in the interphase, expressed per cm^2 of area, can be defined as

$$\left(n_2^{\sigma} - \frac{n_1^{\sigma}}{n_1^{\circ}} \cdot n_2^{\circ} \right) \Big/ A^{\sigma} = \Gamma_{2(1)} \tag{6.95}$$

The symbol (capital gamma) is conventionally used, with a subscript to represent that the excess of component 2 is defined with respect to component 1. For dilute solutions, since component 1, the solvent, is in vast excess, it can be assumed that it has no significant 'surface excess' of either sign. The symbol Γ_2 is then used for the solute.

It follows from equations (6.94) and (6.95) that

$$\Gamma_2 = -\frac{d\gamma}{d\mu_2} \tag{6.96}$$

which is the Gibbs adsorption equation, or the Gibbs adsorption isotherm. In terms of activity of solute, it is

$$\Gamma_2 = -\frac{1}{RT} \frac{d\gamma}{d(\ln a_2)} \tag{6.97}$$

A basic equation of surface chemistry, it expresses quantitatively what is obvious qualitatively: that any solute that reduces surface tension is preferentially 'adsorbed' at the surface. Although hardly necessary, a satisfactory test of the equation was made by slicing off the surface of a solution by a blade travelling at 35 ft s^{-1}. The equation applies to 'negative adsorption' such as shown by aqueous solutions of electrolytes;[25] it is intelligible that ions should shun the interface between water and air—air being a medium in which they cannot seek the satisfaction of their intense fields.

6.7.2   Electrochemical potential

There is electrical dissymmetry at any junction between dissimilar phases. It is impossible to ignore electrical effects at interfaces. If one phase is an ionic solution, the ions must respond to the local, atypical conditions at an interface—one example has just been quoted. An electric field may emanate from a foreign phase in contact with such a solution. It may arise in a variety of ways: an intrinsically polar or self-ionising surface layer, specific adsorption of one kind of ion from solution, an externally imposed electric charge, even oriented adsorption of polar solvent molecules. The question is, whatever the origin of such a field, how will the ions in the solution phase respond to it?

It can be seen qualitatively that positive and negative ions will respond oppositely; attracted or repelled, their populations near the interface will be increased or decreased according to their sign of charge and the direction of the field. Excluding the possibility of discharge of the ions, they will assume an equilibrium distribution tending to annul the field, thus establishing what is traditionally known as an *electrical double layer* on the solution side of the interface. This movement to minimise energy by relieving the stress of the field is impeded by random thermal motion, which maintains the energy–entropy balance appropriate to the prevailing temperature. The field therefore dies away asymptotically towards the bulk of the solution phase, and the double layer is diffuse in this direction. These briefly sketched effects are basic to colloid science, all electrokinetic effects and the behaviour of all systems with aqueous solutions in contact with other phases. Their importance in life-sciences need hardly be stressed.

Fundamental studies in this field require controllable model systems. One such is mercury (phase α) in contact with aqueous sodium chloride solution (phase β). The interfacial tension, γ, can be measured, comparatively but accurately (capillary electrometer), as a function of the electrical polarisation of the mercury, effected by means of a potentiometer and an electrode reversible to chloride ions situated in a remote region of the aqueous solution. The mercury-solution interface approximates closely to the *ideal polarised electrode*, i.e. an electrode that passes no current over a wide range of polarisations. No electrolytic process, with transfer of charge across the metal-solution interface, has any significant tendency to occur. The concentration of every charged constituent is finite in one phase *only.*[26] Such an electrode is nevertheless an equilibrium system; interfacial tension and ionic distribution on the solution side of the interface are self-adjusting in response to polarisation as the independent variable. If this can be accepted, with somewhat less than adequate exposition of *why*, attention can be turned to the thermodynamic treatment of the system, using the Gibbs adsorption isotherm.

The concept of the chemical potential of component i, $\mu_i = \bar{G}_i$, cannot, unmodified, be applied to a charged species situated in a field, or potential gradient. Instead, the *electrochemical potential* must be used. For species i, of charge z_i, situated at a point in phase β where the electrical potential is ψ^β, it is

$$\bar{\mu}_i^\beta = \mu_i^\beta + z_i F \psi^\beta \tag{6.98}$$

where F is the value of the Faraday, the magnitude of the charge carried by a mole of univalent ($z_i = +1$ or -1) ions. The symbol, $\bar{\mu}_i^\beta$, is unlikely to be misleading in the context in which it is used. It is seen that $\bar{\mu}_i^\beta = \mu_i^\beta$ for $z_i = 0$ or $\psi^\beta = 0$. The equilibrium distribution of the species i is such that $\bar{\mu}_1^\beta$ is everywhere the same.

In these terms, the Gibbs equation for the mercury (α)-solution (β) *interphase* is

$$\Sigma \Gamma_i^\alpha d\bar{\mu}_i^\alpha + \Sigma \Gamma_i^\beta d\bar{\mu}_i^\beta + d\gamma = 0 \tag{6.99}$$

The 'components' of the mercury, in which there can be no potential gradient, can be considered as Hg^+ ions and electrons, e^-. In the solution there are Na^+ and Cl^- ions and, for generality, H_2O molecules. Expanding the equation,

$$\Gamma_{Hg^+}^\alpha d\bar{\mu}_{Hg^+}^\alpha + \Gamma_{e^-}^\alpha d\bar{\mu}_{e^-}^\alpha + \Gamma_{Na^+}^\beta d\bar{\mu}_{Na^+}^\beta + \Gamma_{Cl^-}^\beta d\bar{\mu}_{Cl^-}^\beta +$$
$$+ \Gamma_{H_2O}^\beta d\mu_{H_2O}^\beta + d\gamma = 0 \tag{6.100}$$

Each electrochemical potential can be split thus,

$$d\bar{\mu}_{Hg^+}^\alpha = d\mu_{Hg^+}^\alpha + F d\psi^\alpha; (z_i = +1)$$

$$d\bar{\mu}_{e^-}^\alpha = d\mu_{e^-}^\alpha - F d\psi^\alpha; (z_i = -1)$$

$$d\bar{\mu}_{Na^+}^\beta = d\mu_{Na^+}^\beta + F d\psi^\beta; (z_i = +1)$$

and so on. But, for the solution phase, ψ^β is not uniform (as ψ^α must be for the mercury phase). Near the interface, it varies with distance very rapidly, then decreases asymptotically to zero in the homogeneous bulk of the solution phase. For every species, $\bar{\mu}_i^\beta$ is everywhere the same within the phase, and is therefore equal to μ_i^β in the bulk of the solution where ψ^β = 0. Equation (6.100) can therefore be rewritten:

$$\Gamma^\alpha_{Hg^+}d\mu^\alpha_{Hg^+} + \Gamma^\alpha_{Hg^+}Fd\psi^\alpha + \Gamma^\alpha_{e^-}d\mu^\alpha_{e^-} - \Gamma^\alpha_{e^-}Fd\psi^\alpha +$$

$$+ \Gamma^\beta_{Na^+}d\mu^\beta_{Na^+} + \Gamma^\beta_{Cl^-}d\mu^\beta_{Cl^-} + \Gamma^\beta_{H_2O}d\mu^\beta_{H_2O} + d\gamma = 0$$

Imposing the condition of constant composition allows all the $d\mu_i$ terms to be set at zero, leaving

$$(\Gamma^\alpha_{Hg^+} - \Gamma^\alpha_{e^-})Fd\psi^\alpha + d\gamma = 0$$

It is seen that the term in brackets, times F, is the net positive charge, q^α, on the mercury, so that

$$\left(\frac{\partial \gamma}{\partial \psi^\alpha}\right)_{T,P,\ composition} = -q^\alpha \qquad (6.101)$$

which is the *Lippmann equation*; it indicates that as the interfacial tension passes through a maximum along the polarisation scale, the charge on the mercury is zero. This is the *electrocapillary maximum* (ecm). Elsewhere, it allows the charge to be measured and related to polarisation potential in terms of the appropriate slope.

Differentiation of the Lippmann equation with respect to ψ^α:

$$\frac{\partial^2 \gamma}{\partial \psi^{\alpha 2}} = -\frac{\partial q^\alpha}{\partial \psi^\alpha} = -C \qquad (6.102)$$

gives C, called the *differential capacity of the electrical double layer*. This provides an important alternative route to studies of the double layer. In so far as C is constant—which it is not—it can be seen that equation (6.102) requires γ to show a parabolic dependence on ψ^α. In practice, the electrocapillary curves, $\gamma(\psi^\alpha)$, approximate to parabolae, with maxima at the ecm, or point of zero charge.

Readers will agree that in the endeavour to illustrate a principle there is danger of being carried away into too much detail—a halt must now be called to further development of the present theme, except to present a result. In the hands of D. C. Grahame (whose review[26] still presents one of the best brief accounts of the subject), as further development of the argument would show, it became practicable to find the surface excesses of the individual ionic species at the mercury-solution interface. The results for 0·3 molar sodium chloride solution are illustrated in Fig. 6.9.

Looking first at polarisations negative to the ecm (i.e. q^α increasingly negative), it is seen that adsorption of sodium ions follows a rising curve, as to be expected. Chloride ions are repelled in the same region. At polarisations positive to the point of zero charge, chloride ion adsorption rises

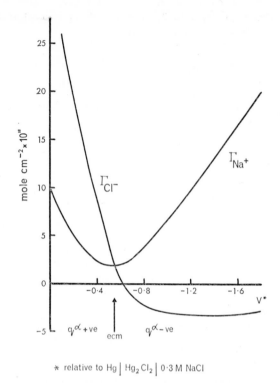

Fig. 6.9   Surface excesses of sodium and chloride ion at the mercury-0·3M aqueous
NaCl solution interface as functions of polarisation.

very steeply, but the remarkable thing is that sodium ions are also attracted—
in effect readsorbed. In addition, it is seen that at the ecm itself, there is a
not very large but significant equal adsorption of both kinds of ion. The
equality is of course necessary because the interphase as a whole must remain
electrically neutral, and here q^α is zero. These observations tell a lot about
the electrical double layer. Briefly, the story is as follows.

It is necessary to recognise two kinds of adsorption of ions. One is a function of the charge on the ion concerned and the charge on the metallic
substrate. It is a function of electrostatic attraction and is called Volta
adsorption. The other is specific adsorption; it is an attraction, of a 'chemical'
nature, not directly dependent on charge, but tending to be strongly
reinforced when Volta adsorption also comes into play. The situation
is very like Brønsted specific, and Debye–Hückel, non-specific interaction
of ions (cf. section 6.4.2.7). In this particular system, chloride ions are
specifically adsorbed. Appreciable negative polarisation of the mercury

is required to expel them from the surface. At positive polarisations, the effect increases rapidly, and *super-equivalent adsorption* of chloride ions occurs—that is to say, far more than necessary to balance the positive charge on the mercury. Since, despite their intimate association with the mercury, the chloride ions retain their charge, they provide an outward field that attracts sodium ions. In this range of polarisation, therefore, the ionic double layer has a negative inner zone and a positive, diffuse outer one. On the opposite side of the ecm, both inner and outer zones are of the same sign—positive.

A comparison of halide ions brings in chemical interest. Specific adsorption, believed zero for F^-, increases rapidly in the sequence Cl^-, Br^-, I^-, yet, lest covalent binding should be too readily assumed, ionic charge is retained. Close to the metal-solution interface, in fields as strong as 10^6 V cm^{-1}, it is to be expected that the properties of water might be greatly changed—by dielectric saturation (near-rigid alignment of molecules) and the compression (~ 1000 atm) attributable to electrostriction. The very great abnormalities in such regions can be illustrated by reference to one of the simplest systems: mercury in contact with 0·01 molar aqueous sodium fluoride solution, at 25 °C. With the mercury negatively polarised at 0·48 V with respect to the ecm, the local 'concentrations' of sodium and fluoride ions at the distance of closest approach are 1·90 mole l^{-1} and 5·3 × 10^{-5} mole l^{-1} respectively. Yet the electrochemical potentials of these species are each uniform throughout the solution phase, and so, of course, is the chemical potential of the electrically neutral *salt*, sodium fluoride; μ_{NaF} $= \mu_{Na^+} + \mu_{F^-}$, the terms $z_i F \psi^\beta$ cancelling.

6.8 Conclusion

It may be helpful to bring the threads of a long chapter together by looking to see whether our ideas on the chemical potential of a component need clarifying. We seize on to the concept that, provided there is no impediment to the transfer of the components of a system between any parts of the system, they will reach a distribution such that μ_i, for each component will be everywhere the same. This remains true whether the system is in equilibrium in all other respects, or in disequilibrium in some. The criterion is that for the transfer of dn_i mole of component i from any place to any other place, *without any other change whatsoever*, $(dG)_{T,P} = 0$. We have seen circumstances in which it is strictly necessary to exclude any other change accompanying the transfer, such as $(dG)_{T,P,n_i,A} = 0$. If we do not do this, we are testing for equilibrium in other respects than the distribution of the components in the system as it stands. If this is what we want to do, it may be necessary to bring in other significant potentials. Suppose that the gravitational field is significant—as it certainly is for the enormously compressible state of matter just above the critical temperature. $\mu_i^\alpha + M_i\phi^\alpha = \mu_i^\beta + M_i\phi^\beta$ is the condition for equilibrium between levels α and β, M_i being mass and ϕ gravitational potential. This is precisely the kind of operation used (taking liberties with the definition of component) to deal with charged species, as we have seen. Temperature and pressure are of course familiar potentials, linking with μ_i and others in the determination of the energy–entropy balance at equilibrium.

REFERENCES

1. Rowlinson, J. S., *Liquids and Liquid Mixtures*, Butterworths, London, 1959.
2. Ives, D. J. G., and Lemon, T. H., *Roy. Inst. Chem. Rev.* 1968, **1**, 62; Eisenberg, D., and Kauzmann, W., *The Structure and Properties of Water*, Oxford, 1969.
3. Franks, F., and Johnson, H. H., *Trans. Faraday Soc.*, 1962, **58**, 656.
4. Franks, F., and Ives, D. J. G., *Quart. Rev.*, 1966, **20**, 1.
5. Bernal, J. D., *Proc. Roy. Soc.*, 1964, **A**, **280**, 299; *Scient. American*, no. 267, August, 1960.
6. Eyring, H., and Marchi, R. P., *J. Chem. Educ.*, 1963, **40**, 562; Eyring, H., and Mu Shik Jhon, *Significant Liquid Structures*, John Wiley, 1969.
7. Smith, E. B., *Intermolecular Forces*, Ann. Reports, 1966, **63**, 13; Parsonage, N. G., *Gases, Liquids and Liquid Mixtures*, ibid., 1967, **64**, A, 57; 1968, **65**, A, 33.
8. Hildebrand, J. H., and Scott, R. L., *Solubility of Non-electrolytes*, 3rd ed., Reinhold, 1950; Dover, 1964.
9. Beatty, H. A., and Callingaert, L., *Ind. Eng. Chem.*, 1934, **26**, 904.
10. McGlashan, M. L., *J. Chem. Educ.*, 1963, **40**, 516.
11. Rice, O. K., *Statistical Mechanics, Thermodynamics and Kinetics*, Freeman, 1967.
12. Lewis, G. N., and Randall, M., *Thermodynamics*, 1923, chap. XVII, McGraw-Hill.
13. *Symbols, Signs, and Abbreviations recommended for British Scientific Publications*, Symbols Committee of the Royal Society, 1969.
14. Ives, D. J. G., *Chem. Brit.*, 1969, **5**, 522.
15. MacInnes, D. A., *The Principles of Electrochemistry*, Reinhold, 1939.
16. Lewis, G. N., and Randall, M., *Thermodynamics*, p. 286, McGraw-Hill, 1923.
17. Robinson, R. A., and Stokes, R. H., *Electrolyte Solutions*, Butterworths, 1959.
18. Pitzer, K. S., and Brewer, L., *Thermodynamics*, McGraw-Hill, 1961.
19. Brønsted, J. N., and la Mer, V. K., *J. Amer. Chem. Soc.*, 1924, **46**, 555.
20. Franks, F., *Physico-chemical Processes in mixed aqueous Solvents*, Heinemann, 1967; *Hydrogen-bonded Solvent Systems*, Taylor and Francis, 1968.

21. Davenport, A. J., Rowlinson, J. S., and Saville, G., *Trans. Faraday Soc.*, 1966, **62**, 322.
22. Prigogine, I., *The Molecular Theory of Solutions*, North-Holland, 1957.
23. Discussions of the Faraday Society, 1953, **15**.
24. Francis, A. W., *Critical Solution Temperatures*, Advances in Chemistry Series no. 31, 1961, American Chemical Society.
25. Randles, J. E. B., *Advances in Electrochemistry and Electrochemical Engineering*, vol. 3, p. 1, John Wiley, 1963.
26. Grahame, D. C., *Chem. Rev.*, 1947, **41**, 441.

Appendix 1

It becomes increasingly difficult to exclude statistical mechanics from discussion, in order to keep within agreed confines. To avoid frustration, the following semi-tabular statement is offered to form a link with other studies.

Classical statistics	Quantum statistics	
Boltzmann	Fermi–Dirac	Bose–Einstein

It is a matter of permutations and combinations to arrive at the following formulae for W, the number of ways n particles can be distributed over p states; it is easy to check them with pencil and paper using small numbers.

$$W = \frac{n!}{n_1! n_2! \dots n_i!}$$	$$W = \frac{p!}{n!(p-n)!}$$	$$W = \frac{(n+p-1)!}{n!(p-1)!}$$
the denominator to include terms for all states. Particles distinguishable. Arrangement within states not significant, exchange between states significant.	$p \geqslant n$ (0! = 1)	
	Particles indistinguishable	
	$\not> 1$ particle per state	No restriction on no. of particles per state

Although it is basic that all microstates of varying W are 'impartially explored' by a macroscopic system, the assumption is justified that the steady equilibrium state of a system large enough not to show fluctuations can be calculated on the basis that there is no significant departure from the distribution of particles over states that maximises W. Within the restrictions of constant total number of particles and constant total energy, maximisation of W leads to the following distributions:

$$n_i = \frac{g_i}{\exp(\alpha + \beta \varepsilon_i)}$$	$$n_i = \frac{g_i}{\exp(\alpha + \beta \varepsilon_i) + 1}$$	$$n_i = \frac{g_i}{\exp(\alpha + \beta \varepsilon_i) - 1}$$

where n_i is the number of particles in the ith state of energy ε_i and degeneracy g_i,

$$e^{-\alpha} = \frac{n}{\Sigma g_i \exp(-\varepsilon_i/kT)} \quad \text{and} \quad \beta = \frac{1}{kT}.$$

Statistical mechanics is normally developed from Boltzmann statistics (seen to be satisfactory when $\exp(\alpha + \beta\varepsilon_i) \gg 1$), using the distribution law

$$\frac{n_i}{n} = \frac{g_i \exp(-\varepsilon_i/kT)}{\Sigma g_i \exp(-\varepsilon_i/kT)}$$

of which the denominator is the *partition function*. This, however, in terms of $S = k \ln W$, overestimates the translational entropy of a gas by $k \ln N!$ per mole, where N is the Avogadro number. This is because distinguishability is postulated where there is none, and correction is necessary.

Appendix 2

The mole fraction of solute, x_2, molality, m_2 and molar concentration,* C_2, of a given solution are related by

$$x_2 = \frac{m_2}{m_2 + (1000/M_1)} = \frac{C_2}{C_2 + \{(1000 - C_2 M_2)/M_1\}}$$

$$= \frac{m_2 M_1}{m_2 M_1 + 1000} = \frac{C_2 M_1}{C_2(M_1 - M_2) + 1000\rho} \tag{A2.1}$$

where M_1 and M_2 are molecular weights of solvent and solute respectively, and ρ is the density of the solution in g ml^{-1} or g cm^{-3} as the case may be.*

For a given solution, the chemical potential of the solute,

$$\mu_2 = \left(\frac{\partial G}{\partial n_2}\right)_{T,P,n_1},$$

is independent of what composition scale, or standard state, is chosen. Hence,

$$\mu_2 = \mu_{2(x)}^\circ + RT \ln \gamma_{2(x)} x_2 = \mu_{2(m)}^\circ + RT \ln \gamma_{2(m)} m_2$$

$$= \mu_{2(C)}^\circ + RT \ln \gamma_{2(C)} C_2 \tag{A2.2}$$

Consider a second solution, of the same solute in the same solvent, of such extreme dilution that it is ideal, and the activity coefficient of the solute is unity on whatever scale it is expressed. For this solution, indicating the great dilution by the subscript ∞:

$$\mu_{2(\infty)} = \mu_{2(x)}^\circ + RT \ln x_{2(\infty)} = \mu_{2(m)}^\circ + RT \ln m_{2(\infty)}$$

$$= \mu_{2(C)}^\circ + RT \ln C_{2(\infty)} \tag{A2.3}$$

Clearly, $\mu_2 - \mu_{2(\infty)}$ is independent of composition scales, so that

$$\mu_2 - \mu_{2(\infty)} = RT \ln \gamma_{2(x)} \cdot \frac{x_2}{x_{2(\infty)}} = RT \ln \gamma_{2(m)} \cdot \frac{m_2}{m_{2(\infty)}}$$

$$= RT \ln \gamma_{2(C)} \cdot \frac{C_2}{C_{2(\infty)}} \tag{A2.4}$$

* The litre, once 1000 cm³, was used to establish the kg as the mass of 1 litre of water at its temperature of maximum density. The prototype kg became the primary standard, and more precise measurements required restandardisation of the litre. It became 1000·028 cm³. The recent recommendation is to redefine it as 1000 cm³ exactly. The change, 0·003% is not normally significant, but, when it is, the new proposal will be troublesome, and the view can be taken that it would be better to leave well alone.

Then $\gamma_{2(x)} \cdot \dfrac{x_2}{x_{2(\infty)}} = \gamma_{2(m)} \cdot \dfrac{m_2}{m_{2(\infty)}} = \gamma_{2(C)} \cdot \dfrac{C_2}{C_{2(\infty)}}$ (A2.5)

For the excessively dilute solution equation (A2.1) in the limit becomes

$$x_{2(\infty)} = \frac{m_{2(\infty)}M_1}{1000} = \frac{C_2 M_1}{1000\rho_\infty}$$ (A2.6)

where ρ_∞ may be identified with the density of the solvent.
Substitution from equations (A2.1) and (A2.6) into (A2.5) leads to

$$\gamma_{2(x)} = \gamma_{2(m)}\left(1 + \frac{m_2 M_1}{1000}\right)$$ (A2.7)

$$\gamma_{2(x)} = \gamma_{2(C)}\left\{\frac{\rho}{\rho_\infty} + C_2(M_1 - M_2)/1000\rho_\infty\right\}$$ (A2.8)

$$\gamma_{2(m)} = \gamma_{2(C)}\left\{\frac{\rho}{\rho_\infty} - C_2 M_2/1000\rho_\infty\right\}$$ (A2.9)

Index